ARCHITECTS CONTRACTORS ENGINEERS

Guide To Construction COSTS

2005 Vol. XXXVI

Architects Contractors Engineers Guide to Construction Costs 2005, Volume 36.

ISBN 1-58855-067-2

ARCHITECTS CONTRACTORS & ENGINEERS
GUIDE TO CONSTRUCTION COSTS

EDITOR'S NOTE 2005

This annually published book is designed to give a uniform estimating and cost control system to the General Building Contractor. It contains a complete system to be used with or without computers. It also contains Quick Estimating sections for preliminary conceptual budget estimates by Architects, Engineers and Contractors. Square Foot Estimating is also included for preliminary estimates.

The Metropolitan Area concept is also used and gives the cost modifiers to use for the variations between Metropolitan Areas. This encompasses over 50% of the industry. This book is published annually to be historically accurate with the traditional May-July wage contract settlements and to be a true construction year estimating and cost guide.

The Rate of Inflation in the Construction Industry in 2004 was *8%*. Labor contributed a *4%* increase and materials rose *10%*.

The Wage Rate for Skilled Trades increased an average of *4%* in 2004. Wage rates will probably increase at a *4%* average next year.

The Material Rate increased *10%* in 2004. The main increases were in lumber, steel, piping, steel products, plywood and copper products.

Construction Volume should stay level 2005. Housing will probably continue to rise, and Industrial and Commercial Construction should stay flat. Highway and Heavy Construction should decrease.

The Construction Industry had its largest inflation in 12 years. Some Materials should inflate at a slower pace, and Labor should again inflate slightly in 2005.

We are recommending using a *5%* increase in your estimates for work beyond July 1, 2004.

Don Roth
Editor

State	Metropolitan Area	Multiplier
AK	ANCHORAGE	137
AL	ANNISTON	90
	AUBURN-OPELIKA	90
	BIRMINGHAM	88
	DOTHAN	86
	GADSDEN	86
	HUNTSVILLE	88
	MOBILE	89
	MONTGOMERY	86
	TUSCALOOSA	88
AR	FAYETTEVILLE-SPRINGDALE-ROGERS	80
	FORT SMITH	86
	JONESBORO	85
	LITTLE ROCK-NORTH LITTLE ROCK	87
	PINE BLUFF	87
	TEXARKANA	86
AZ	FLAGSTAFF	98
	PHOENIX-MESA	98
	TUCSON	97
	YUMA	99
CA	BAKERSFIELD	112
	FRESNO	114
	LOS ANGELES-LONG BEACH	115
	MODESTO	111
	OAKLAND	120
	ORANGE COUNTY	114

State	Metropolitan Area	Multiplier
CA	REDDING	111
	RIVERSIDE-SAN BERNARDINO	112
	SACRAMENTO	114
	SALINAS	116
	SAN DIEGO	113
	SAN FRANCISCO	125
	SAN JOSE	122
	SAN LUIS OBISPO	110
	SANTA CRUZ-WATSONVILLE	116
	SANTA ROSA	117
	STOCKTON-LODI	113
	VALLEJO-FAIRFIELD-NAPA	116
	VENTURA	112
	SANTA BARBARA	115
CO	BOULDER-LONGMONT	96
	COLORADO SPRINGS	101
	DENVER	101
	FORT COLLINS-LOVELAND	94
	GRAND JUNCTION	96
	GREELEY	94
	PUEBLO	98
CT	BRIDGEPORT	112
	DANBURY	112
	HARTFORD	111
	NEW HAVEN-MERIDEN	112
	NEW LONDON-NORWICH	109

State	Metropolitan Area	Multiplier
CT	STAMFORD-NORWALK	115
	WATERBURY	111
DC	WASHINGTON	104
DE	DOVER	104
	WILMINGTON-NEWARK	105
FL	DAYTONA BEACH	90
	FORT LAUDERDALE	94
	FORT MYERS-CAPE CORAL	86
	FORT PIERCE-PORT ST. LUCIE	93
	FORT WALTON BEACH	95
	GAINESVILLE	88
	JACKSONVILLE	92
	LAKELAND-WINTER HAVEN	88
	MELBOURNE-TITUSVILLE-PALM BAY	96
	MIAMI	94
	NAPLES	96
	OCALA	93
	ORLANDO	93
	PANAMA CITY	81
	PENSACOLA	86
	SARASOTA-BRADENTON	87
	TALLAHASSEE	83
	TAMPA-ST. PETERSBURG-CLEARWATER	91
	WEST PALM BEACH-BOCA RATON	94
GA	ALBANY	85
	ATHENS	87
	ATLANTA	95
	AUGUSTA	83
	COLUMBUS	84
	MACON	87
	SAVANNAH	87
HI	HONOLULU	135
IA	CEDAR RAPIDS	97
	DAVENPORT	100
	DES MOINES	101
	DUBUQUE	95
	IOWA CITY	100
	SIOUX CITY	95
	WATERLOO-CEDAR FALLS	94
ID	BOISE CITY	99
	POCATELLO	97
IL	BLOOMINGTON-NORMAL	106
	CHAMPAIGN-URBANA	105
	CHICAGO	115
	DECATUR	104
	KANKAKEE	107
	PEORIA-PEKIN	106
	ROCKFORD	106
	SPRINGFIELD	104
IN	BLOOMINGTON	101
	EVANSVILLE	100
	FORT WAYNE	100
	GARY	108
	INDIANAPOLIS	104
	KOKOMO	100
	LAFAYETTE	100
	MUNCIE	100
	SOUTH BEND	101
	TERRE HAUTE	100
KS	KANSAS CITY	99
	LAWRENCE	94
KS	TOPEKA	93
	WICHITA	93
KY	LEXINGTON	93
	LOUISVILLE	96
	OWENSBORO	94
LA	ALEXANDRIA	88
	BATON ROUGE	92
	HOUMA	92
	LAFAYETTE	90
	LAKE CHARLES	92
	MONROE	88
	NEW ORLEANS	95
	SHREVEPORT-BOSSIER CITY	89
MA	BARNSTABLE-YARMOUTH	115
	BOSTON	118
	BROCKTON	114
	FITCHBURG-LEOMINSTER	111
	LAWRENCE	114
	LOWELL	109
	NEW BEDFORD	114
	PITTSFIELD	110
	SPRINGFIELD	110
	WORCESTER	109
MD	BALTIMORE	98
	CUMBERLAND	96
	HAGERSTOWN	91
ME	BANGOR	95
	LEWISTON-AUBURN	96
	PORTLAND	97
MI	ANN ARBOR	106
	DETROIT	111
	FLINT	105
	GRAND RAPIDS-MUSKEGON-HOLLAND	100
	JACKSON	104
	KALAMAZOO-BATTLE CREEK	96
	LANSING-EAST LANSING	104
	SAGINAW-BAY CITY-MIDLAND	102
MN	DULUTH	106
	MINNEAPOLIS-ST. PAUL	111
	ROCHESTER	106
	ST. CLOUD	108
MO	COLUMBIA	99
	JOPLIN	95
	KANSAS CITY	102
	SPRINGFIELD	97
	ST. JOSEPH	99
	ST. LOUIS	99
MS	BILOXI-GULFPORT-PASCAGOULA	87
	JACKSON	86
MT	BILLINGS	98
	GREAT FALLS	99
	MISSOULA	97
NC	ASHEVILLE	82
	CHARLOTTE	83
	FAYETTEVILLE	84
	GREENSBORO-WINSTON-SALEM-H POINT	83
	GREENVILLE	83
	HICKORY-MORGANTON-LENOIR	79
	RALEIGH-DURHAM-CHAPEL HILL	83
	ROCKY MOUNT	79
	WILMINGTON	84

State	Metropolitan Area	Multiplier
ND	BISMARCK	95
	FARGO	96
	GRAND FORKS	94
NE	LINCOLN	92
	OMAHA	96
NH	MANCHESTER	100
	NASHUA	98
	PORTSMOUTH	92
NJ	ATLANTIC-CAPE MAY	110
	BERGEN-PASSAIC	113
	JERSEY CITY	112
	MIDDLESEX-SOMERSET-HUNTERDON	108
	MONMOUTH-OCEAN	111
	NEWARK	114
	TRENTON	111
	VINELAND-MILLVILLE-BRIDGETON	109
NM	ALBUQUERQUE	96
	LAS CRUCES	92
	SANTA FE	98
NV	LAS VEGAS	108
	RENO	107
NY	ALBANY-SCHENECTADY-TROY	102
	BINGHAMTON	100
	BUFFALO-NIAGARA FALLS	108
	ELMIRA	95
	GLENS FALLS	94
	JAMESTOWN	102
	NASSAU-SUFFOLK	115
	NEW YORK	132
	ROCHESTER	106
	SYRACUSE	103
	UTICA-ROME	103
OH	AKRON	103
	CANTON-MASSILLON	101
	CINCINNATI	98
	CLEVELAND-LORAIN-ELYRIA	106
	COLUMBUS	101
	DAYTON-SPRINGFIELD	99
	LIMA	101
	MANSFIELD	97
	STEUBENVILLE	104
	TOLEDO	103
	YOUNGSTOWN-WARREN	100
OK	ENID	88
	LAWTON	89
	OKLAHOMA CITY	91
	TULSA	90
OR	EUGENE-SPRINGFIELD	106
	MEDFORD-ASHLAND	104
	PORTLAND	108
	SALEM	107
PA	ALLENTOWN-BETHLEHEM-EASTON	105
	ALTOONA	103
	ERIE	103
	HARRISBURG-LEBANON-CARLISLE	101
	JOHNSTOWN	104
	LANCASTER	101
	PHILADELPHIA	114
	PITTSBURGH	104
	READING	103
	SCRANTON-WILKES-BARRE-HAZLETON	102

State	Metropolitan Area	Multiplier
PA	STATE COLLEGE	98
	WILLIAMSPORT	100
	YORK	102
RI	PROVIDENCE	109
SC	AIKEN	91
	CHARLESTON-NORTH CHARLESTON	85
	COLUMBIA	85
	FLORENCE	82
	GREENVILLE-SPARTANBURG-ANDERSON	84
	MYRTLE BEACH	90
SD	RAPID CITY	89
	SIOUX FALLS	90
TN	CHATTANOOGA	88
	JACKSON	87
	JOHNSON CITY	84
	KNOXVILLE	88
	MEMPHIS	93
	NASHVILLE	92
TX	ABILENE	86
	AMARILLO	87
	AUSTIN-SAN MARCOS	88
	BEAUMONT-PORT ARTHUR	88
	BROWNSVILLE-HARLINGEN-SAN BENITO	89
	BRYAN-COLLEGE STATION	87
	CORPUS CHRISTI	85
	DALLAS	91
	EL PASO	85
	FORT WORTH-ARLINGTON	91
	GALVESTON-TEXAS CITY	90
	HOUSTON	92
	LAREDO	78
	LONGVIEW-MARSHALL	83
	LUBBOCK	88
	MCALLEN-EDINBURG-MISSION	84
	ODESSA-MIDLAND	85
	SAN ANGELO	83
	SAN ANTONIO	89
	TEXARKANA	86
	TYLER	86
	VICTORIA	86
	WACO	86
	WICHITA FALLS	86
UT	PROVO-OREM	94
	SALT LAKE CITY-OGDEN	93
VA	CHARLOTTESVILLE	88
	LYNCHBURG	87
	NORFOLK-VA BEACH-NEWPORT NEWS	91
	RICHMOND-PETERSBURG	91
	ROANOKE	85
VT	BURLINGTON	98
WA	BELLINGHAM	112
	BREMERTON	110
	OLYMPIA	108
	RICHLAND-KENNEWICK-PASCO	106
	SEATTLE-BELLEVUE-EVERETT	111
	SPOKANE	108
	TACOMA	111
	YAKIMA	106
WI	APPLETON-OSHKOSH-NEENAH	101
	EAU CLAIRE	100
	GREEN BAY	101

State	Metropolitan Area	Multiplier
WI	JANESVILLE-BELOIT	101
	KENOSHA	103
	LA CROSSE	99
	MADISON	100
	MILWAUKEE-WAUKESHA	107
	RACINE	103
	WAUSAU	100

State	Metropolitan Area	Multiplier
WV	CHARLESTON	99
	HUNTINGTON	100
	PARKERSBURG	98
	WHEELING	100
WY	CASPER	92
	CHEYENNE	93

HOW TO USE THIS BOOK

Labor Unit Columns

- Units *include* all taxable (vacation) and untaxable (welfare and pension) fringe benefits.
- Units *do not include* taxes and insurance on labor (approximately 35% added to labor cost). (Workers Compensation, Unemployment Compensation, and Employers FICA.)
- Units *do not include* general conditions and equipment (approx. 10%).
- Units *do not include* contractors' overhead and profit (approx. 10%).
- Units are Union or Government Minimum Area wages.

Material Unit Columns

- Units *do not include* general conditions and equipment (approx. 10%).
- Units *do not include* sales or use taxes (approx. 5%) of material cost.
- Units *do not include* contractors' overhead and profit (approx. 10%).
- Units are mean prices FOB job site.

Cost Unit Columns (Subcontractors)

- Cost is complete with labor, material, taxes, insurance and fees of a subcontractor. This also applies to Quick Estimating sections.
- Units *do not include* general contractors' overhead or profit.

Quick Estimating Sections - For Preliminary and Conceptual Estimating

- Includes all labor, material, general conditions, equipment, taxes, insurance and fees of contractors combined for items not covered in Cost Unit columns.

Construction Modifiers - For Metro Area Cost Variation Adjustments

- Labor items are percentage increases or decreases from 100% for labor units used in the book's labor unit columns. (See first page of each Division.)
- Material and Labor combined are percentage increases and decreases from 100% for metro variations used in the book's Unit Cost columns, Quick Estimating sections and Sq.Ft. Cost Section - see Page 2.
- Recommend Unit Costs of 10% less than Book Units for Residential Construction.

EXAMPLE:	UNIT	LABOR	MATERIAL	COST
2" x 4" Wood Studs & Plates - 8' @ 16" O.C.	BdFt	.67	.55	1.22
Add 35% Taxes & Insurance on Labor		.23	-	.23
Add 5% Material Sales Tax		-	.03	.03
				1.48
Add 10% General Conditions & Equipment				.15
				1.63
Add 10% Overhead & Profit				.17
BdFt Cost - Bid Type Estimating			BdFt	1.80
or SqFt Wall Cost with 3 Plates - Quick Estimating (page 6A-14)			or SqFt	1.40
See Division 4 for Other Examples (Masonry)				

Purchasing Practice (usual) in the industry for each section of work is indicated immediately after section heading in parenthesis and defined as follows:

- L&M - Labor and Material Subcontractor - Firm bid to G.C. according to plans and specifications for all labor and material in that section or subsection.
- M - Material Subcontractor - Firm bid to G.C. according to plans and specifications for all material in that section or subsection.
- L - Labor Subcontractor - Firm bid to G.C. to *install only* material furnished according to plans and specifications by others.
- S - Material Supplier - Firm Price for materials delivered only.
- E - Equipment Supplier - Firm Price for Equipment Rented or Purchased.

Installation Practice by particular trades are listed behind each section heading and are according to "Agreement and Decisions Rendered Affecting the Building Industry," by the A.F.L. Building and Construction Trades Department, A.F.L.-C.I.O. Area variations, jurisdictional problems and multiple jurisdictions are possible.

DIVISION #1 - GENERAL REQUIREMENTS

General Requirements amount to approximately 6% to 15% of total job costs for General Construction and 5% to 10% for Residential Construction. All items below are included in G.R. except 0101.32, .33, and .34 (taxes and insurance on labor). See Square Foot Costs (SF Section) for more accurate Dollar Value of Job Costs.

			UNIT	LABOR	MATERIAL
0101.0	**GENERAL CONDITIONS - Specified (AIA Doc A201)**				
.1	SUPERINTENDENT Job Size/Complexity		Week	$1,800	-
.2	PERMITS AND FEES				
	Concrete and Commercial	- Approx. Average	MCuFt	-	$3.50
	Residential (Finish Area)	- Approx. Average	SqFt	-	$ 1.00
	Remodeling	- Approx. Average	M$	-	$6.00
	Varies with Each Permit Division				
.3	TAXES (Art. 3.6)				
.31	Sales and Use Tax (Varies as to State)		%		Avg. 5.7%

		Payroll Max	Per $100 or % Gross Payroll	
.32	Unemployment Compensation	to $ 22,000	State Maximum	9.1%
	(Varies as to State)		State Minimum	.7%
.33	Unemployment Compensation	to $ 7,000	Fed. Maximum	.8%
.34	FICA - Employer's Share	to $ 87,900		6.2%
	Medicare	No Limit		1.45%
.4	INSURANCE - CONTRACTOR'S LIABILITY			
	(Worker's Compensation, P.L. & P.D.)			
	Classification Avg - Contractors Experience		Gross Payroll	14.0%
	Clerical - Office	(#8810)	" "	1.0%
	Supervision	(#5606)	" "	2.5%
	Excavation	(#6217)	" "	12.0%
	Concrete - Structural - N.O.C.	(#5213)	" "	20.0%
	Slabs on Ground, Flat	(#5221)	" "	8.0%
	Masonry	(#5022)	" "	12.0%
	Iron and Steel - N.O.C.	(#5057)	" "	32.0%
	Carpentry - N.O.C. Commercial	(#5403)	" "	19.0%
	Interior & Dwellings	(#5437)	" "	12.0%
	Wallboard	(#5445)	" "	13.0%
	Plumbing			7.0%
	Electrical			6.0%
	Wrecking (Demolition)			40.0%
	PL. and PD.			1.5%
	Add for Assigned Risk			75%

		UNIT	LABOR	MATERIAL
.5	INSURANCE - HOLD HARMLESS			
	(By Owner) (Art. 11.2) (Contract Amount)	M$		$ .80
.6	INSURANCE - ALL RISK (Fire/Ext.Cov./Theft/etc.)			
	(By Owner) (Art. 11.2) ($1,000 Deductible)			
	Fire Proof - All Concrete	Annual)		$2.20
	Fire Resistive - Covered Steel	Insurable)		$2.75
	Ordinary - Masonry, Ext. Walls & Joists	Value)		$3.80
	Frame, Wood	per M$)		$6.50
	Deduct for $2,500 Deductible			10%
	Add for Light Protection Areas - Approx.			50%
.7	CLEANUP (Art. 3.15)			
.71	Progress	Bldg - SqFt	$.18	$ .05
.72	Glass, 2 Sides	Glass Area - SqFt	$.35	-
.8	PERFORMANCE & PAYMENT BOND	(If Specified) (Art. 11.4)		
	$ -0- to $ 99,999	Job Bureau Comp. Contr. Amount		2.50%
	$ 100,000 to $ 499,999			1.50%
	$ 500,000 to $2,499,999			1.00%
	$2,500,000 to $4,999,999			.75%
	$5,000,000 to $7,499,999			.70%
	$7,500,000 Up			.65%
.9	SAFETY OF PERSONS & PROPERTY - Job Condition			
.10	TESTS AND SAMPLES - See Appropriate Division			

0102.0 SPECIAL CONDITIONS (Changes to G.C. 0101 and Temporary Utilities & Facilities, usually Specified)

		UNIT	LABOR	MATERIAL
.1	TEMPORARY OFFICE			
	Fixed - 10' x 16" Wood	Each	1,450	2,500
	Mobile (Special Construction) 8' x 32'	Each	-	4,000
		Month	-	200
	Mobile (Special Construction) 10' x 50'	Each	1,100	7,000
		Month	-	300
	Mobile (Special Construction) 12' x 60'	Each	1,300	10,000
		Month	-	400
	Add for Delivery & Pickup	Each	-	400
.2	TEMPORARY STORAGE SHED -12' x 24'	Each	1,800	3,500
.3	TEMPORARY SANITARY FACILITIES - Chemical Toilet	Month	-	100
.4	TEMPORARY TELEPHONE (No Long Distance) - Per line	Month	-	95
	TEMPORARY TELEPHONE	Initial	-	300
	TEMPORARY TELEPHONE - Cellular	Month	-	125
	TEMPORARY FAX	Initial	-	1,200
.5	TEMPORARY POWER AND LIGHT			
	Initial Service 110V	Job	-	1,250
	220V	Job	-	1,900
	440V	Job	-	3,500
	Monthly Service and Material	Month	-	250
	Temporary Wiring and Lighting-Total Job	SqFt	-	.08
.6	TEMPORARY WATER -Initial Service	Initial	Job Condition	
	Monthly Service	Month	-	100.00
.7	TEMPORARY HEAT (after Enclosed)			
	See Divisions 3 & 4 for Temporary Heat, Concrete & Masonry			
	See Division 1A for Equipment Rental Rates			
	Heaters:			
	Steam & Unit - Heater Rent - SqFt Htd Area	Month	-	.12
	Piping - SqFt Htd Area	Month	-	.14
	Condensate - SqFt Htd Area	Month	-	.19
	Mobile Home Heater Rent - SqFt Htd Area	Month	-	.12
	Fuel SqFt Htd Area	Month	-	.25
	Tending 8 - Hour/Day	8 Hrs	-	240.00
.8	TEMPORARY ENCLOSURES AND COVERS			
	Per Door - Wood	Each	32.00	45.00
	Per Window - Polyethylene and Frame	Each	14.00	12.00
	Walls - Frame and Reinforced Poly - 1 use	SqFt	.22	.18
	2 uses	SqFt	.20	.14
	Frame and 1/2" Plywood - 1 use	SqFt	.60	1.00
	2 uses	SqFt	.50	.65
	Floors- 2" Fiberglass & 7 mil Poly- 1 use	SqFt	.12	.45
.9	TEMPORARY ROADS & RAMPS - No Surfacing	SqYd	Job Condition	
.10	TEMPORARY DEWATERING - No Tending			
	Pumping - 2" Pump	Day		95.00
	3" Pump	Day	-	105.00
	4" Pump	Day	-	125.00
	Add for Tending			
	Well Points - LnFt Header	Month	-	105.00
	Add Each Month	Month	-	80.00
.11	TEMPORARY SIGNS - 4' x 8'	Each	100.00	450.00
.12	TEMP. FENCING, BARRICADES & WARNING LIGHTS			
	6' Chain Link - Rented (Erected and Removed)	LnFt	1.25	5.50
	Add for Barbed Wire	LnFt	.35	1.00
	Add for Gates	LnFt	8.00	20.00
	Snow Fence - New	LnFt	.40	1.25
	Concrete Barriers (Rented - Month)	LnFt	-	4.00
	Flasher Warning Light (Rented - Month)	Each	-	20.00
	8' Barricade (Rented - Month)	Each	-	45.00
	12' Barricade (Rented - Month)	Each	-	55.00

		UNIT	LABOR*	MATERIAL
0103.0	**MISCELLANEOUS CONDITIONS (Non-Specified)**			
.1	FIELD ENGINEERING AND LAYOUT	Week		Job Cond.
.2	TIMEKEEPERS AND ACCOUNTANTS	Week		Job Cond.
.3	SECURITY	Week		Job Cond.
.4	OPERATING ENGINEERS FOR HOISTING	Week		Job Cond.
.5	TRAVEL AND SUBSISTENCE - Field Employees			Job Cond.
.6	TRAVEL AND SUBSISTENCE - Office Employees			Job Cond.
.7	OFFICE SUPPLIES AND EQUIPMENT	Week		25.00
.8	COMMUNICATIONS EQUIPMENT			
	Radio-Multi Channel Hand Unit with Charger & Battery	Each	-	1,350
	Base Unit - F.M. with Antenna and Mike	Each	-	1,500
.9	HOISTING EQUIPMENT (Towers, Cranes and Lifts)			
.91	Towers and Hoists - Portable Self-Erecting Style			
	Skip (Based on 100' Unit) -3000# to 5000# Capacity	Month	-	2,300
	2000# to 3000# Capacity	Month	-	2,100
	Add to Heights Above 100'	LnFtMo	-	21
	Towers - Fixed (based on 100') - 30' Override			
	Tubular Frame - Double Well - 5000# Capacity	Month	-	2,000
	Installation - Double Well - Subcont - Up	LnFt	65	-
	Double Well - Down	LnFt	50	-
	Add for Concrete Bucket and Hopper	Month	-	400
	Add for Chicago Boom	Month	-	230
	Add for Gates, Signals, etc (Safety)	Job	750	1,200
	Hoist Engine, Gas Powered - 2 Drum	Month	-	1,800
	Towers and Hoists - Personnel 6000# Capacity			
	Based on 100' Unit - 8 1/2' Override	Month	-	3,500
	Add Above 100'	LnFt	-	35
	Installation - Up	LnFt	-	40
	Down	LnFt	-	38
.92	Cranes - Climbing - 20' Override			
	Based on 125' Bldg-130' Boom, 4000#	Month	-	6,500
	to 200' Boom, 7000#	Month	-	9,000
	230' Boom, 7000#	Month	-	10,500
	Add for LnFt Tower above 125'	Month	-	40
	Installation			
	Foundation	Each	-	12,000
	440-Volt Transformer Install	Each	-	2,500
	Installation - Up & Down	LnFt	-	110
	Up	Each	-	15,000
	Down	Each	-	15,000
	Avg - Up, Down & Haul - 125' Tower	Each	-	30,000
	Add per Lift - Jacking	LnFt	-	50
	Add per Floor Pattern	Each	-	1,000
.93	Cranes - Mobile Lifting (no Operator)	HOUR	DAY	MONTH
	15-Ton Hydraulic	110	600	3,800
	20-Ton	120	650	4,000
	25-Ton	125	700	6,600
	30-Ton	140	750	7,000
	45-Ton	160	800	7,800
	75-Ton Cable	175	900	8,700
	90-Ton	200	1,100	9,500
	125-Ton	240	1,500	13,000
.94	Fork Lifts - Construction	UNIT	LABOR	MATERIAL
	Four-Wheel Drive - 28' 5000# Capacity	Month		2,300
	36' 6000# Capacity	Month		2,400
	37' 6000# Capacity	Month		2,500
	42' 6000# Capacity	Month		2,600
	42' 8000# Capacity	Month		2,700

Add for Local Travel & Setup Time - 4 Hours

*Subcontract

0103.0 MISCELLANEOUS CONDITIONS, Cont'd...

			UNIT	LABOR	MATERIAL
.95	Platform Lifts				
	Scissor Lifts – Electric	26'	Month		615
		32'	Month		800
	Telescoping Boom Lifts	46'	Month		4,800
		66'	Month		2,450
	Articulating Boom Lifts	86'	Month		5,100
		36'	Month		1,500
		51'	Month		1,900
		66'	Month		2,450
.10	TRUCKS AND OTHER JOB VEHICLES (Site Use)				
	$ 500,000 Job		JobAvg		1,700
	$1,000,000 Job		JobAvg		6,200
	$5,000,000 Job		JobAvg		14,500
	1-Ton Pickup		Month		625
	2-Ton Flatbed		Month		880
	2 1/2-Ton Dump		Month		1,500
.11	MAJOR EQUIPMENT (See Div. 2,3,4,5,& 6 if Subbed)				
	(See 1-6A & 7A: Rental Rates, Purchase Prices)				
	$ 100,000 Job		JobAvg		4,000
	$ 500,000 Job		JobAvg		7,800
	$1,000,000 Job		JobAvg		14,500
	$5,000,000 Job		JobAvg		58,000
	As a % of Labor (No Hoisting Equipment)				5%
.12	SMALL TOOLS AND SUPPLIES				
	(See 1-6A & 7A: Rental Rates, Purchase Prices)				
	$ 100,000 Job		JobAvg		1,400
	$ 500,000 Job		JobAvg		3,800
	$1,000,000 Job		JobAvg		6,500
	$5,000,000 Job		JobAvg		30,000
	As a Percentage of Labor		JobAvg		2%
.13	HAULING EQUIPMENT AND TOOLS				
	$ 100,000 Job		JobAvg	950	600
	$ 500,000 Job		JobAvg	1,800	1,500
	$1,000,000 Job		JobAvg	3,500	3,000
	$5,000,000 Job		JobAvg	11,000	8,000
.14	GAS AND OIL FOR EQUIPMENT				
	$ 500,000 Job		JobAvg		1,200
	$1,000,000 Job		JobAvg		1,700
	$5,000,000 Job		JobAvg		7,500
	As a Average of Equipment Cost		JobAvg		12%
.15	REPAIRS TO EQUIPMENT AND TOOLS				
	$ 100,000 Job		JobAvg		600
	$ 500,000 Job		JobAvg		1,200
	$1,000,000 Job		JobAvg		2,200
	$5,000,000 Job		JobAvg		7,200
	As an Average of Equipment Cost				10%
.16	PHOTOGRAPHS - 2 Views, 4 Copies		Month		140
.17	SCHEDULING - C.P.M., Pert, Network, Analysis				
	$ 500,000 Job		JobAvg		4,500
	$1,000,000 Job		JobAvg		8,000
	$5,000,000 Job		JobAvg		14,000
	Add for Update, Monitor		Each		750
.18	COMPUTER COSTS		Employee/Wk		10
.19	PERFORMANCE BOND - SUBCONTRACTORS		ContrAmt		1.5%
.20	PUNCH LIST AND FINAL OUT		ContrAmt		.5%

TOOLS & EQUIPMENT

See 103.9 for Towers, Cranes and Lift Equipment	Approximate Cost New	Fair Rental* Rate per Mo.
Div. 1 General Conditions		
Engineering Instruments - Transits	1,229	250
Levels 18"	998	230
Fans - 15,000 CFM - 36"	1,155	400
Generators - 3500 Watt - Portable	1,260	375
5000 Watt - Portable	1,628	425
Heaters, Temp- Oil 100M to 300 BTU	1,365	450
LP Gas 90M BTU	378	200
150M BTU	420	240
350M BTU	746	350
Laser - Rotating	4,800	600
Light Stands - Portable - 500 watt Quartz	704	260
Pumps- Submersible 1/2 HP Elect 2"	441	320
Centrifugal 2" 8,000 Gal	1,019	490
3" 17,000 Gal	1,050	550
Diaphragm 2" - 3" x 3" - Pneumatic	1,470	460
3" - 3" x 3" - Gas	1,523	590
Submersible - 5 HP 3"	1,575	500
Hose - Discharge - 3" - 25'	105	55
Suction - 3" - 20'	137	70
Vacuum Cleaners - Wet Dry - 5 Gal	630	160
Div. 2 Excavation (Including Demolition)		
Air Tools- Breakers Light Duty - 35 to 40#	1,008	260
Heavy Duty - 60 to 90#	1,386	340
Rock Drill - 40#	1,869	350
Chipping Hammer - 7# to 12#	809	275
Clay Digger	756	250
Hose - 3/4" - 50'	137	60
Compressors 125 cfm Gas - Tow	11,235	800
185 cfm Diesel - Trailer Mounted	14,280	825
26' Conveyor - 2 HP Electric	9,608	980
Electric Breakers - 60#	2,205	410
Jacks - Hydraulic - 20 Ton - 11 HP	840	320
Screw - 20 Ton	378	85
Saws - Chain 18"	578	395
Concrete Asphalt 14" Blade	1,680	250
24" Plate - Vibratory Gas 170#	2,205	645
Tampers - Rammer Gas 135#	2,415	740
Div. 3 Concrete		
Bucket & Boot- 1 CuYd	2,153	480
1 1/2 CuYd	2,730	490
2 CuYd	3,570	750
Buggies - Manual - 6 to 8 CuFt	494	180
Power Operated - 16 CuFt	6,300	850
Conveyors - 32' Hydraulic and Belt - Gas	9,870	1,220
50' Hydraulic and Belt - Gas	21,840	2,625
Core Drill - Diamond - 15 Amp	3,413	550
with Trailer	9,870	900
Crack Chaser	4,410	1,350
Drill - Cordless 3/8" Reversing	305	100
Heavy Duty 1/2" Reversing	389	150
3/4" Reversing	704	195
Sander- Hand - 9"	347	150
Grinder- Ceiling 7' to 11' - Elec	2,310	440
Floor - Elec - 1 1/2 HP	3,203	650
Hammer - Drill - 3/4" Variable Speed	494	190
Roto Chipping - 1 1/2"	525	490
Hopper - Floor 30 CuFt	1,155	230
Mixer - 6 CuFt	2,415	480
Planer - 10'	4,515	720
Scarifier & Scabblers - Single Bit	1,155	725
5 Bit Floor - 175 cfm	8,190	2,200
Stripper - Floor	3,938	430

*Weekly Rent 40% of monthly, Daily Rent 15% of monthly, Hourly Rent 2% of monthly.

TOOLS & EQUIPMENT

		Approximate Cost New	Fair Rental* Rate per Mo.
Div. 3	**Concrete, Cont'd...**		
	Saws - Rotary Hand - 7 1/4"	190	130
	9 1/4" - 8 1/4"	235	140
	Table 14" Radial Arm 3 HP	1,800	300
	Floor - Gas 11 HP	1,550	650
	Gas 18 HP	4,300	690
	Electric 8 HP 14"	1,820	650
	Cut off 14" Electric	710	475
	Screeds - Vibrating 6'	900	675
	12 1/2' Beam Type	1,950	685
	25' Beam Type	4,200	780
	Troweling Machines - Gas 36" 4 Blade 7 HP	1,700	500
	48" 4 Blade 6 HP	1,950	560
	Trunks & Tremmies - 12" and Hopper	180	65
	Vibrators - Flexible Head - 1 HP	1,100	330
	2 HP	1,200	360
	High cycle - 180 Cycle	3,200	550
	Back Panel 2 1/2 HP	750	675
	Wheelbarrows	115	60
Div. 4	**Masonry**		
	Hoist Electric - One ton	1,050	170
	Mixer - Mortar - 6 Cu Ft - Gas 7 HP	2,000	430
	Elec 1 1/2 HP	2,120	400
	8 Cu Ft - Gas 7 HP	2,400	450
	12 Cu Ft - Gas or Electric	3,600	940
	Pallet Jack	740	300
	Pump Grout - Hydraulic 10 GPM	6,200	1,400
	Saws - Table - 1 1/2 HP - 14"	2,150	400
	5 HP - 14"	3,100	450
	Scaffold - 5' - 6" with Braces	75	3.60
	6' - 6" with Braces	80	4.00
	Wheels	85	11.00
	Brackets	22	2.00
	Plank - Aluminum	90	5.00
	16' Laminated	35	3.00
	Base - Adj	25	1.60
	Splitter - 10,000#	1,700	380
	Rolling Towers	350	100
	Swing Stage - Motorized with deck-cables-safety	9,000	1,200
Div. 6	**Carpentry**		
	Door Hanging Eq. (Router, Mortiser, Template)	300	100
	Drills - Heavy Duty (See Div. 3)		100
	Fasteners - Power - Drywall	150	45
	Hammer - Drill 1/2"	310	170
	Miter Box - 14" Power	410	330
	Nailers and Staplers - Spot with compressor	1,800	280
	Planers	500	150
	Sanders - Disc 9"	260	100
	Belt 3" - Heavy Duty	300	125
	Floor	420	330
	Saws - Band	300	250
	Chain	350	330
	Rotary, Hand - 7 1/4"	190	95
	8 1/4"	220	130
	Table - Fixed 10"	400	140
	Jig	200	80
	Reciprocating	450	220
	Cut-off - 12" Elec.	2,000	375
	Scaffold - Rolling	450	110
	Screwdriver - Variable Speed - Drywall	250	110
	Stud Drivers - Low Velocity	280	120
	High Velocity	400	140
	Wrench - 1/2" Impact - Air	290	140
	1" Impact - Air	600	485

*Weekly Rent 40% of monthly, Daily Rent 15% of monthly, Hourly Rent 2% of monthly.

See Division 1501 for Piped Utility Material (CSI 02600)
See Division 1501 for Water Distribution (CSI 02660)
See Division 1501 for Fuel Distribution (CSI 02680)
See Division 1503 for Sewage Distribution (CSI 02700)
See Division 1600 for Power Communication (CSI 02780)

		UNIT	COST with MACHINE
0201.0	**DEMOLITION & CLEARING (Op.Eng., Truck Dr. & Lab.)**		
.1	STRUCTURE MOVING - Wood Frame	SqFt	14.00
	With Masonry	SqFt	17.00
.2	CLEARING AND GRUBBING		
	Trees - Trunk	Dia/Inch	14.00
	Stump - Removed	Dia/Inch	11.00
	Chipped - Below Grade	Dia/Inch	6.25
	Add for Hauling Away and Dumping	Dia/Inch	5.20
	Shrubs & Brush - Light	Acre	2,100.00
	Medium	Acre	2,900.00
	Heavy	Acre	3,900.00
.3	DEMOLITION AND LOAD (See .34 for Disposal)		
.31	Total Building		
	Wood - Mainly Housing	CuFt/Bldg	.30
	Masonry - Load Bearing	CuFt/Bldg	.35
	Concrete - Frame	CuFt/Bldg	.44
	Steel - Frame	CuFt/Bldg	.35
	Add for Floor Above 2 Stories	CuFt/Bldg	.06

.32 Selective Building Removals (No Cutting/Disposal)
(See .35 - Cut and Drill)
(See .34 - Disposal/Haul Away)
(See .33 - Site Removal)

	Unit	Labor Only	Machine Only	Unit	Labor Only	Machine Only
Concrete:						
8" Walls Reinforced	CuYd	257.00	94.00	SqFt	6.60	1.85
Non-Reinforced	CuYd	203.00	83.00	SqFt	5.05	1.60
12" Walls Reinforced	CuYd	350.00	73.00	SqFt	12.80	2.75
Footings - 24" x 12"	CuYd	278.00	161.00	LnFt	10.50	6.00
6" Struc. Slab Reinforced	CuYd	220.00	65.00	SqFt	4.10	1.16
8" Struc. Slab Reinforced	CuYd	233.00	86.00	SqFt	5.80	1.70
4" Slab on Ground Reinforced	CuYd	415.00	54.00	SqFt	1.90	.63
Non-Reinforced	CuYd	375.00	47.00	SqFt	1.45	.60
6" Slab on Ground Reinforced	CuYd	236.00	44.00	SqFt	2.30	.80
Non-Reinforced	CuYd	225.00	41.00	SqFt	1.65	.76
9" Stairs Reinforced	CuYd	296.00	93.00	SqFt	7.05	2.60
Masonry:						
4" Brick or Stone Walls	CuYd	110.00	32.00	SqFt	1.30	.40
and 8" Backup	CuYd	198.00	68.00	SqFt	2.30	.70
Block or Tile Partitions	CuYd	84.00	30.00	SqFt	1.05	.36
6" Block or Tile Partitions	CuYd	68.00	23.00	SqFt	1.18	.44
8" Block or Tile Partitions	CuYd	60.00	42.00	SqFt	1.39	.53
12" Block or Tile Partitions	CuYd	158.00	57.00	SqFt	1.79	.65
Add for Plastered Type				SqFt	.40	.12

Misc - Hand Work (Add for Equipment & Scaffold)	UNIT	Labor	
Acoustical Ceilings - Attached (Incl. Iron)	SqFt	.60	-
Suspended (Incl. Grid)	SqFt	.38	-
Asbestos - Ceilings and Walls	SqFt	15.20	-
Columns and Beams	SqFt	37.20	-
Pipe	LnFt	46.50	-
Tile Flooring	SqFt	1.75	-
Cabinets and Tops	LnFt	10.50	-
Carpet	SqFt	.32	
Ceramic and Quarry Tile	SqFt	1.20	-
Doors and Frames - Metal	Each	53.00	-
Wood	Each	47.25	-
Drywall Ceilings - Attached (Including Grid)	SqFt	.75	-
Wood or Metal Studs - 2 Sides	SqFt	.81	-
Paint Removal - Doors and Windows	SqFt	.75	-
Walls	SqFt	.59	-
Plaster on Wood or Metal Studs	SqFt	1.10	-
Ceilings - Attached (Incl. Iron)	SqFt	1.05	-
Roofing - Built-Up	SqFt	1.00	-
Shingles - Asphalt and Wood	SqFt	.34	-
Terrazzo Flooring	SqFt	1.75	-
Vinyl Composition Flooring (No Mastic Removal)	SqFt	.36	-
Wall Coverings	SqFt	.65	-
Windows - Metal or Wood	Each	35.00	-
Wood Flooring	SqFt	.37	-

0201.0 DEMOLITION & CLEARING, Cont'd...

		UNIT	(1) LABOR ONLY	MACHINE ONLY
.33	Site Removals			
	(See .34 for Disposal)			
	(See .35 for Cutting Additions)			
	Concrete:4" Sidewalks - Reinforced	SqFt	2.00	.65
	Non -Reinforced	SqFt	1.57	.62
	6" Drives - Reinforced	SqFt	2.35	.83
	Non-Reinforced	SqFt	1.80	.87
	Curb - 6" x 18"	LnFt	5.45	1.55
	and Gutter	LnFt	8.75	2.07
	Asphalt:2"	SqFt	1.20	.40
	Curb	LnFt	2.80	.70
	Fencing: 8" Metal	LnFt	1.80	.44
.34	Disposal or Haul Away (Including Truck or Box & Driver)			
	Haul - No Dump Charges Included			COST
	Concrete Truck Measure - 75¢ a minute or	CuYd/Mile		.78
	Masonry - 75¢ a minute or	CuYd/Mile		.75
	Wood - 75¢ a minute or	CuYd/Mile		.70
	Box Rent (Included Above) - 30 Cu.Yd.	Box		375.00
	Clean Material - 20 Cu.Yd.	Box		340.00
	10 Cu.Yd.	Box		290.00
	Dump Charges - Concrete	CuYd		7.10
	Building Materials	CuYd		10.30
	Tree and Brush Removal	CuYd		14.45
.35	Saw Cutting/Core Drilling/Torch Cutting/Blasting			
	Saw Cutting (includes Operator):			
	Slabs - Asphalt	LnFt	per inch	1.17
	Concrete - Structural	LnFt	per inch	2.15
	Hollow	LnFt	per inch	.90
	Slabs on Grd/Non - Rein.	LnFt	per inch	.78
	/with Mesh	LnFt	per inch	1.00
	Walls - Concrete - Reinforced	LnFt	per inch	3.35
	Non - Reinforced	LnFt	per inch	2.67
	Masonry - Hollow	LnFt	per inch	1.93
	Solid	LnFt	per inch	2.50
	Core Drilling:			
	Solid Slab - 2" Diameter	Inch		4.40
	3" Diameter	Inch		5.00
	4" Diameter	Inch		6.65
	6" Diameter	Inch		9.30
	8" Diameter	Inch		12.40
	Hollow Core Slab - 2" Diameter	Inch		2.70
	3" Diameter	Inch		2.85
	4" Diameter	Inch		4.15
	6" Diameter	Inch		7.60
	Torch Cutting - Steel per Inch	LnFt		5.30
	Blasting - Unrestricted	CuYd		115.00
	Restricted	CuYd		205.00
	Cut Openings:			
	Concrete (see .32)	SqFt		-
	Sheet Rock Partitions	SqFt		3.00
	Wood Floors	SqFt		2.50
	No Disposal or Haul Away Included (see .34)			

(1) With Pneumatic Tools

0202.0 EARTHWORK (Op. Eng., Truck Driver & Lab.)

.1 GRADING

Machine	UNIT (1)	COST	UNIT (1)	COST
Strip Top Soil - 4"	CuYd	5.13	SqYd	.57
Spread Top Soil				
4" Site Borrow	CuYd	5.67	SqYd	.63
Off-Site Borrow ($13 CuYd)	CuYd	21.60	SqYd	2.40

Rough Grading - Cut and Fill	UNIT	SOFT (1)	MED. (1)	HARD (1)
Dozer - To 200'	CuYd	3.50	3.80	4.30
To 500'	CuYd	4.00	5.60	6.95
Scraper - Self Propelled - To 500'	CuYd	3.20	3.55	4.15
To 1,000'	CuYd	4.00	4.70	5.50
Grader - To 200'	CuYd	3.50	4.05	5.90
Add for Compaction (See 0202.4)				
Add for Hauling Away (See 0202.5)				
Add for Truck Haul on Site	CuYd	2.80	-	-
Add for Borrow Brought In				
Loose Fill – 3.00 Ton	CuYd	11.10	-	-
Crushed Stone - 5.50 Ton	CuYd	14.60	-	-
Gravel - 4.30 Ton	CuYd	12.90	-	-

Hand and Machine	UNIT	HAND LABOR	MACHINE
Fine Grading - 4" Fill - Site	CuYd	12.15	6.50
	or SqFt	.15	.08
4" Fill - Building	CuYd	21.90	10.50
	or SqFt	.27	.13
.2 EXCAVATION - Dig & Cast or Load (No Hauling)			
Soil			
Open - Soft (Sand) with Backhoe or Shovel	CuYd	22.00	3.40
Medium (Clay)	CuYd	30.00	4.15
Hard	CuYd	45.20	5.85
Add for Clamshell or Dragline	CuYd	-	1.35
Deduct for Front End Loader	CuYd	-	1.15
Deduct for Dozer	CuYd	-	.28
Trench or Pocket - Soft with Backhoe or Shovel	CuYd	23.60	5.00
Medium	CuYd	30.30	6.50
Hard	CuYd	43.25	7.80
Add for Clamshell or Dragline	CuYd	-	1.25
Augered - 12" diameter	CuYd	49.60	9.40
24" diameter	CuYd	38.00	6.35
Add for Frost	CuYd	-	7.50
Rock - Soft - Dozer	CuYd	-	33.20
Hammer	CuYd	-	82.00
Medium - Blast	CuYd	-	100.00
Hammer	CuYd	-	166.00
Hard - Blast	CuYd	-	220.00
Hammer	CuYd	-	295.00
Swamp	CuYd	-	9.40
Underwater	CuYd	-	12.30
.3 BACKFILL (Not Compacted and Site Borrow)	CuYd	10.60	2.70
Add for Off-Site Borrow	CuYd	-	11.00
Add to <u>All Above</u> for Mobilization Average			5%

(1) Machine and Operators

		UNIT	LABOR (1)	COST (2)
0202.0	**EARTHWORK, Cont'd...**			
.4	COMPACTION (Incl. Backfilling & Site Borrow)			
	Labor Only			
	Building to 85 - 24" Lifts	CuYd	15.50	-
	to 95 - 12" Lifts	CuYd	18.00	-
	to 100 - 6" Lifts	CuYd	21.50	-
	Labor Compaction & Machine Backfill Combined			
	Building to 85 - 24" Lifts	CuYd	9.30	2.90
	to 90 - 18" Lifts	CuYd	9.80	3.00
	to 95 - 12" Lifts	CuYd	10.50	3.10
	to 100 - 6" Lifts	CuYd	11.50	3.50
	Subgrade (Including Grading and Rollers)	CuYd	-	3.95
	Machine Only (No Grading)			
	Site to 85 - 24" Lifts	CuYd		3.30
	to 90 - 18" Lifts	CuYd		3.40
	to 95 - 12" Lifts	CuYd		3.95
	to 100 - 6" Lifts	CuYd		4.90
	Add to Above for Off-site Borrow	CuYd		11.00
.5	DISPOSAL (Haul Away) Country	CuYd/Mile		.68
	City	CuYd/Mile		.80
.6	TESTS			
	Field Density - Compaction	Each		37.50
	Soil Gradation	Each		54.00
	Density (Proctor)	Each		88.00
	Moisture and Specific Gravities	Each		93.00
	Hydrometer (Sand, Silt and Clay)	Each		104.00
	Rock -2" Core	LnFt		66.50
	4" Core	LnFt		108.00
	Add for Soil Penetration	LnFt		7.50
	Add for Mileage	Mile		.44
0203.0	**PILE AND CAISSON FOUNDATIONS (L&M) (Operating Engineers and Iron Workers)**			
.1	CONCRETE			
.11	Augured or Cast in Place -12"	LnFt		21.50
.12	Precast Prestressed -10"	LnFt		18.80
	12"	LnFt		22.00
	14"	LnFt		24.00
	16"	LnFt		29.00
.13	Pipe Casing Pulled - Cast - in - Place - 10"	LnFt		20.00
	12"	LnFt		21.00
.2	STEEL			
.21	Steel Pipe -10"	LnFt		25.00
	12"	LnFt		29.00
	14"	LnFt		33.50
	16"	LnFt		36.50
	Concrete - Filled - 10"	LnFt		28.00
	12"	LnFt		34.00
	14"	LnFt		40.00
	16"	LnFt		46.00
.22	Steel H Section 8" - 36#	LnFt		23.00
	10" - 42#	LnFt		26.00
	12" - 53#	LnFt		31.00
	14" - 73#	LnFt		33.00
	14" - 89#	LnFt		41.50
.23	Helical Pier Foundation			
	8" -10" - 12"	LnFt		24.00
	10" - 12" - 14"	LnFt		26.00
.3	WOOD (Treated) to 40' (Incl. Tests & Cutoffs)			
	10"	LnFt		17.50
	12"	LnFt		18.80
	14"	LnFt		21.00
	16"	LnFt		22.00
	Add for Mobilization			10%

(1) Hand Work
(2) Machine and Operators

		UNIT	COST
0203.0	**PILE & CAISSON, Cont'd...**		
.4	CAISSONS -20" Concrete Filled	LnFt	33.00
	24"	LnFt	39.00
	30"	LnFt	53.00
	36"	LnFt	67.00
	42"	LnFt	86.00
	48"	LnFt	108.00
	54"	LnFt	145.00
	60"	LnFt	175.00
	Add for Bells	Each	1,850.00
	Add for Wet Ground	LnFt	5.50
	Add for Obstructions - Hand Removal	CuFt	60.00
	Add for Obstructions - Machine Removal	CuFt	11.00
	Add for Reinforced Steel	Ton	1,600.0
	Add for Mobilization		10%
0204.0	**EARTH & WATER RETAINER WORK**		
.1	STEEL SHAFT PILING (Including Bracing) (L&M)		
	Heavy -27# -12"-16' Full Salvage	SqFt	17.50
	40# -15"-16' Full Salvage	SqFt	18.50
	57# -18"-16' Full Salvage	SqFt	19.50
	Add per Foot Longer than 16'	SqFt	2.00
	Add for No Salvage	SqFt	8.00
	Light -12 ga - 18" Black to 8' Full Salvage	SqFt	14.50
	10 ga - 18" Black to 8' Full Salvage	SqFt	15.60
	8 ga - 18" Black to 8' Full Salvage	SqFt	19.50
	Add for No Salvage	SqFt	7.25
	Add for Galvanized	SqFt	30%
.2	WOOD SHEET PILING & BRACING		
	6' Deep	SqFt	5.50
	8' Deep	SqFt	5.75
	10' Deep	SqFt	6.30
	12' Deep	SqFt	6.75
.3	H PILES & WOOD SHEATHING (4" Lbr.)		
	10' Deep	SqFt	21.00
	15' Deep	SqFt	22.00
	20' Deep	SqFt	27.00
	25' Deep	SqFt	31.00
	Deduct for Salvage	SqFt	40%
.4	UNDERPINNING	CuYd	700.00
.5	RIP RAP & LOOSE STONE WALLS		
	Machine Placed	CuYd	23.00
	Hand Placed	CuYd	55.00
	and Grouted	CuYd	135.00
.6	CRIBBING		
	6" Open to 10'	SqFt	23.00
	8" Open	SqFt	30.00
0205.0	**SITE DRAINAGE (Including Building Foundations)**		
.1	DITCHING AND BACKFILL (Open) - See 0202.2		
.2	DEWATERING		
	Pumping -2" Pump	Day	57.00
	3" Pump	Day	87.00
	4" Pump	Day	115.00
	Well Points - Ln.Ft. Header - first month	Month	120.00
	Add for Each Additional Month	Month	68.00
	Add for Attending Time	Hour	45.00

0205.0 SITE DRAINAGE, Cont'd...

		UNIT	LABOR	MATERIAL
.3	CULVERTS & PIPING (No Excavation - See 0202.2)			
.31	Corrugated Galvanized Metal			
	Watertight (Bituminous Coated)			
	12" Diameter with Bands & Gaskets	LnFt	4.40	8.00
	15"	LnFt	4.70	12.00
	18"	LnFt	5.25	15.00
	24"	LnFt	6.15	20.00
	30"	LnFt	7.35	25.00
	36"	LnFt	8.70	31.00
	48"	LnFt	17.50	48.50
	Non - Watertight			
	12" Diameter with Bands & Gaskets	LnFt	3.50	6.80
	15"	LnFt	4.25	7.50
	18"	LnFt	5.00	10.00
	24"	LnFt	7.00	12.60
	30"	LnFt	8.90	15.00
	36"	LnFt	10.30	25.10
	48"	LnFt	11.00	34.50
	Add for Oval Arch Shape	LnFt	20%	-
	Aprons			
	18"	Each	17.40	66.00
	24"	Each	19.80	98.00
	30"	Each	22.00	160.00
	36"	Each	25.00	256.00
.32	Reinforced Concrete Pipe - #3 - Class 5			
	12" Diameter	LnFt	5.70	11.00
	18"	LnFt	7.75	17.00
	24"	LnFt	11.00	23.00
	27"	LnFt	13.20	35.00
	30"	LnFt	16.70	42.00
	36"	LnFt	20.50	46.00
	48"	LnFt	31.00	72.00
	Add for Each Added Foot in Depth	LnFt	5%	-
	Add for Apron 36" (Flared End)	Each	170.00	480
	Add for Trash Guards (Galvanized)	Each	145.00	820
.33	Vitrified Clay 12 "	LnFt	6.00	12.00
	24"	LnFt	10.00	45.00
.4	CATCH BASINS & MANHOLES (No Excavation)			
	48" Cast in Place (Concrete)	LnFt	152.00	155.00
	48" Brick or Block	LnFt	148.00	160.00
	48" Precast Concrete	LnFt	98.00	93.00
	48" Precast Concrete Collar	LnFt	126.00	160.00
	Add for Cover - Precast Concrete	Each	69.00	145.00
	Add for Base - Precast Concrete	Each	90.00	150.00
	Add for Covers & Grates - H.D.C.I.	Each	41.00	170.00
	Add for Covers & Grates - L.D.C.I.	Each	41.00	170.00
	Add for Covers - H.D. Watertight	Each	41.00	290.00
	Add for Steps	Each	-	15.00
	Add for Adjusting Rings 2"	Each	14.50	15.00
.5	FOUNDATION DRAINAGE (Subdrainage)			
	4" Clay Pipe	LnFt	1.20	1.70
	4" Corrugated - Perforated	LnFt	1.00	1.00
	4" Plastic Pipe - Perforated	LnFt	1.00	1.15
	Solid	LnFt	1.00	.90
	6" Clay Pipe	LnFt	1.45	1.85
	6" Corrugated - Perforated	LnFt	1.15	1.45
	6" Plastic Pipe - Perforated	LnFt	1.15	2.60
	Solid	LnFt	1.10	2.10
	Add for Porous Surround 2'x 2' ($8.00 CuYd)	LnFt	1.45	2.80

0206.0 PAVEMENT, CURBS AND WALKS

.1 PAVING (PARKING LOTS AND DRIVEWAYS)

		UNIT	COST
.11	Bituminous (Material $185/Ton, 5% Asphalt, 10-Mile Delivery)		
	1 1/2" Wearing 10.5 SqYd/Ton	SqYd	5.40
	2" Wearing 9.0 SqYd/Ton	SqYd	6.85
	2 1/2" Wearing 7.5 SqYd/Ton	SqYd	7.80
	3" Wearing 6.0 SqYd/Ton	SqYd	9.05
	Add for Paths and Small Driveways	SqYd	5.00
	Add for Patching	SqYd	11.30
	Add for Asphalt Content per 1%	SqYd	2.30
	Add for Over 10-mile Delivery - Ton/Mile	SqYd	.40
	Deduct for Base Asphalt	SqYd	11%
	Add for Seal Coat	SqYd	1.70
	Add for Striping - Paint	LnFt	.30
	Add for Striping - Plastic	LnFt	.58
	Base Course for Above - Add:		
	4" Sand & Gravel $8.80 Ton and 1/3 Ton	SqYd	4.30
	4" Crushed Stone $10.90 Ton	SqYd	5.10
	6" Sand & Gravel $7.70 Ton and 1/2 Ton	SqYd	5.20
	6" Crushed Stone $10.00 Ton	SqYd	6.25
	Add per Mile over 10 Miles	SqYd	.40
	Deduct for Belly Dump Delivery - $.75 Ton	SqYd	.40
	4" Sand & Gravel	Ton	9.30
	4" Crushed Stone	Ton	10.40
	6" Sand & Gravel	Ton	8.35
	6" Crushed Stone	Ton	9.70
	Add per Ton Mile Over 10 Miles	Ton	.40
	Deduct for Belly Dump Delivery	Ton	.90
.12	Concrete (4000# Ready Mix, $74 CuYd - Machine Placed)		
	6" Reinforced with 6 x 6, 8-8 Mesh	SqYd	23.00
	7" Reinforced with 6 x 6, 6-6 Mesh	SqYd	23.90
	8" Reinforced with #4 Rods - 12" O.C.	SqYd	27.10
	9" Reinforced with #5 Rods - 12" O.C.	SqYd	29.50
	10" Reinforced with #6 Rods - 8" O.C.	SqYd	30.60
	12" Reinforced with #8 Rods - 8" O.C.	SqYd	38.80
	Add for Base - Same as Base Prices Above		
	Add to Above for Hand Placed	SqYd	25%
	Add for Driveways	SqYd	50%
.13	Stabilized Aggregate (Delivery by Dump Truck)		
	4" Gravel - $8.80 Ton - 3/4" Compacted	SqYd	5.10
	4" Crushed Stone - $10.00 Ton - 3/4" Compacted	SqYd	5.90
	6" Gravel - $8.00 Ton - 3/4" Compacted	SqYd	5.85
	6" Crushed Stone - $9.30 Ton - 3/4" Compacted	SqYd	7.50
	6" Pulverized Concrete	SqYd	7.90
	Add for Soil Cement Treatment	SqYd	1.00
	Add for Oil Penetration Treatment	SqYd	3.10
	Add for Calcium Chloride Treatment	SqYd	.52
	Add per CuYd Mile Over 10 Miles	SqYd	.42
	Deduct for Belly Dump Truck Delivery	SqYd	.43
.14	Brick - 4" x 8" x 2 1/2"	SqYd	80.00
	Add for Sand Bed - Compacted	SqYd	6.80
	Add for Mortar Setting - Bed & Joints	SqYd	7.40

0206.0 PAVEMENT, CURBS AND WALKS, Cont'd...

		UNIT	LABOR	MATERIAL
.2	CURBS & GUTTERS (Cost Incl. Excavation & Equip.)			
.21	Concrete - Cast in Place			
	Curb			
	6" x 12" Small Job - 200' Day	LnFt	5.85	3.00
	6" x 18"	LnFt	7.05	3.55
	6" x 24"	LnFt	8.90	4.60
	6" x 30"	LnFt	10.40	5.55
	Add for Reinforced - Two #5 Rods	LnFt	.70	.70
	Add for Gutter - 6" x 12"	LnFt	4.35	2.40
	Add for Gutter - 6" x 18"	LnFt	4.60	3.30
	Add for Large Job - Over 200' Day	LnFt		10%
	Add for Formless 1000' Day Up	LnFt		20%
	Add for Curved or Radius Work	LnFt		40%
	Curb & Gutter Rolled			
	6" x 12" Small Job	LnFt	7.45	4.40
	6" x 18"	LnFt	8.70	6.05
	6" x 24"	LnFt	10.50	7.80
	Deduct for Large Job - Over 200' Day	LnFt	-	10%
	Deduct for Formless 1000' Day Up	LnFt	-	25%
	Add for Reinforced - Two #5 Rods	LnFt	.68	.80
	Add for Curved or Radius Work	LnFt	-	40%
	Add for Approaches - 8' x 16' x 6'	SqFt	3.70	2.30
.22	Concrete - Precast & Pinned - 10" x 10'	LnFt	1.95	7.35
	9" x 10'	LnFt	1.85	6.80
.23	Asphalt - 6" x 8'	LnFt	1.00	1.35
	Add for Curved Work	LnFt	-	30%
.24	Granite - 6" x 16"	LnFt	5.55	24.00
.25	Timbers, Treated - 6" x 6"	LnFt	1.95	3.60
	6" x 8"	LnFt	2.60	5.20
.26	Plastic - 6" x 6"	LnFt	1.80	5.60
.3	WALKS AND DRIVEWAYS			
	Bituminous - 1 1/2" with 4" base	SqFt	.44	.66
	2" with 4" base	SqFt	.50	.69
	Concrete - 4" - Broom Finish	SqFt	1.52	1.25
	5" - Broom Finish	SqFt	1.58	1.50
	6" - Broom Finish	SqFt	1.65	1.55
	Add Mesh 6" x 6", 10 - 10	SqFt	.14	.13
	Add for Exposed Aggregate - Washed	SqFt	.65	.06
	Add for Coloring (Acrylic)	SqFt	.38	.65
	Add for Base - 4" Sand or Gravel	SqFt	.27	.10
	Precast Block - Colored 1" with 2" Sand Cushion	SqFt	1.43	.84
	Colored 2" with 2" Sand Cushion	SqFt	1.62	1.05
	Crushed Rock - 4" Compacted	SqFt	.33	.30
	Flagstone - 1 1/4" with Sand Bed	SqFt	3.90	7.80
	1 1/4" with Mortar Setting Bed	SqFt	5.50	8.80
	Brick - 4" x 8" x 2 1/4" with 4" Sand Bed	SqFt	3.90	3.25
	4" x 8" x 2 1/4" with Mortar Setting Bed	SqFt	4.50	4.40
	Add for Herringbone or Weave Pattern	SqFt	1.35	1.40
	Wood - 2" T&G on 6" x 6" Timbers	SqFt	1.50	2.60
	2" Boards on 4" x 4" Timbers	SqFt	1.35	2.15
	Asphalt Block - 6" x 12" x 3"	SqFt	2.40	4.55
	Slate - 1 1/4"	SqFt	5.80	8.05
.4	STAIRS – EXTERIOR	SqFt	6.85	9.60

0207.0 FENCING (L&M) (Ironworkers or Carpenters)

		UNIT	COST
.1	CHAIN LINK (Galvanized 9 ga) Including Intermediate Posts and Concrete Embedded		
	5' High - 2" Mesh and 2" O.D. Pipe	LnFt	13.00
	6' High	LnFt	14.00
	7' High	LnFt	14.50
	8' High	LnFt	16.50
	10' High	LnFt	31.00
	Add for 1 - or 3 - Strand Barbed Wire	LnFt	2.80
	Add for Vinyl Coated Wire	LnFt	15%
	Add for Aluminum Wire	LnFt	25%
	Add for Wood Slats	LnFt	50%
	Add for Aluminum Slats	LnFt	75%
	Add for Posts - Corner, End and Gate:		
	5' High - 3" O.D.	Each	40.00
	6' High - 3" O.D.	Each	45.00
	7' High - 3" O.D.	Each	57.00
	8' High - 3" O.D.	Each	66.00
	10' High - 3" O.D.	Each	95.00
	Add for Gate Frames:		
	5' High	LnFt	40.00
	6' High	LnFt	42.00
	7' High	LnFt	53.00
	8' High	LnFt	66.00
	10' High	LnFt	80.00
.2	WELDED WIRE (Galvanized 11 ga) Including Intermediate Posts and Concrete Embedded		
	4' High - 2" x 4"	LnFt	10.50
	5' High	LnFt	11.80
	6' High	LnFt	12.80
	Add for Posts:		
	4' Long	Each	23.00
	5' Long	Each	25.00
	6' Long	Each	27.00
	Add for Gate Frames:		
	4' Long	LnFt	22.00
	5' Long	LnFt	27.00
	6' Long	LnFt	29.00
	Add for Slats	LnFt	5.00
.3	MESH WIRE (Galvanized 14 ga)		
	3' High	LnFt	5.25
	4' High	LnFt	6.50
	5' High	LnFt	7.00
.4	WOOD FENCING		
	Rail - 4' Cedar - 3 Boards	LnFt	15.50
	Redwood - 3 Boards	LnFt	16.50
	Pine, Painted - 3 Boards	LnFt	15.00
	Vertical - 6' Pine, Painted	LnFt	18.50
	Redwood	LnFt	23.00
	Cedar	LnFt	23.00
	Weave - 6' Redwood	LnFt	24.00
	8' Redwood	LnFt	28.00
	Picket - 4' Cedar	LnFt	12.00
	6' Cedar	LnFt	15.00
	Split Rail - 4' - 3 Rail	LnFt	8.25
	Add for Post Set in Concrete	Each	5.00
.5	GUARD RAILS - Corrugated - Galvanized with Wood Posts	LnFt	30.00
	Galvanized with Steel Posts	LnFt	32.00
	Wire Cable - with Wood Posts, 3 - Strand	LnFt	15.00
	with Steel Posts, 3 - Strand	LnFt	16.00
	with no posts, 3-strand	LnFt	7.00

0208.0 RECREATIONAL FACILITIES

		UNIT	COST
.1	PLAYING FIELDS AND OUTDOOR COURTS		
.11	Football Fields - Field Size - 300' x 160		
	Minimum Use Size - 306' x 180'		
	Artificial Turf	Each	650,000.00
	Natural Turf	Each	290,000.00
.12	Tennis Courts -Court Size - 78' x 36'		
	Minimum Use Size - 108' x 48'		
	Asphalt with Color Surfacing - 1 Court	Each	24,000.00
	2 Courts	Each	19,500.00
	3 Courts	Each	19,000.00
	Incl. 3" Asphalt, 4" Gravel Base & Striping		
	Synthetic	Each	33,000.00
	Concrete	Each	28,000.00
	Add for Practice Boards	Each	2,500.00
	Add for Net Posts (4" Pipe with Ratchet)	Pair	600.00
	Add for Steel nets	Each	600.00
.13	Track and Field		
	Track - 440 yard - Track Size 21' x 1,320'		
	Synthetic Turf - 1" Polyurethane	Each	140,000.00
	1" Polyurethane	SqYd	40,000.00
	3" Bituminous Base	SqYd	7.50
	6" Gravel Base	SqYd	5.50
	Rubberized Asphalt	Each	65,000.00
	1" with 12% Resiliency	SqYd	11.00
	Add Bituminous Base - same as above	SqYd	7.00
	Cinder	Each	55,000.00
	High Jump - 50' Radius with 12' x 16' P.T.	Each	10,000.00
	Long Jump - 200' x 4' Runway with 20' x 8' P.T.	Each	4,800.00
	Pole Vault - 130' x 4' Runway with 16' x 16' P.T.	Each	3,000.00
.14	Basketball Courts -		
	Concrete - 4" Grade School - Court 42' x 74'	Each	17,000.00
	High School - Court 50' x 84'	Each	20,000.00
	College - Court 50' x 94'	Each	23,500.00
	Asphalt - 1 1/2" - Grade School	Each	8,200.00
	High School	Each	9,200.00
	College	Each	11,100.00
	Add for Back Stops	Each	1,400.00
.15	Volley Ball Courts - Court Size 30' x 60'	Each	10,000.00
	Minimum Use Size - 36' x 66'		
	Concrete - 4"	Each	9,000.00
	Asphalt - 1 1/2"	Each	7,000.00
	Add for Posts and Inserts	Pair	580.00
.16	Shuffleboard Courts - 6' x 52'		
	Concrete - 4"	Each	1,325.00
	Asphalt - 1 1/2"	Each	850.00
.17	Miniature Golf (9 - Hole) Bases - Concrete	Each	10,000.00
	Equipment and Carpet	Each	8,100.00
.18	Horse Shoe Courts - 10' x 40'	Each	560.00
	See Division 11 for Playground Equipment		
.19	Baseball and Softball Fields		
	45' Radius with Backstop	Each	13,000.00
	65' Radius with Backstop	Each	16,500.00

0208.0	RECREATIONAL FACILITIES, Cont'd...	UNIT	COST
.2	RECREATIONAL EQUIPMENT		
	Climbers - Arch	Each	500.00
	Dome	Each	590.00
	Circular	Each	1,050.00
	Catwalks	Each	680.00
	Slides - Straight - 6'	Each	900.00
	8'	Each	1,150.00
	10'	Each	1,700.00
	Spiral - 7'	Each	4,100.00
	10'	Each	6,200.00
	Spring Equipment - Single	Each	390.00
	2 Unit	Each	590.00
	4 Unit	Each	1,080.00
	Swings - 2 Leg - 2 Seat - 8' High	Each	660.00
	2 Leg - 4 Seat	Each	960.00
	3 Leg - 3 Seat - 8' High	Each	740.00
	6 Seat	Each	1,100.00
	Add for 12' High	-	20%
	Add for Nursery Seats	Each	265.00
	Add for Saddle Seats	Each	615.00
	Add for Tire Swings	Each	155.00
	Teeter Totters	Each	980.00
	Whirls - Small - 6'	Each	810.00
	Large - 10'	Each	1,480.00
.3	SITE FURNISHINGS		
	Benches - Steel Leg		
	Stationary - with Black Wood Slats - 8'	Each	400.00
	with Black Aluminum Slats - 8'	Each	535.00
	Portable - 8'	Each	255.00
	Player - No Back, Wood Seat - 8'	Each	210.00
	No Back, Wood Seat - 14'	Each	340.00
	Bike Racks		
	Permanent - 10'	Each	550.00
	20'	Each	850.00
	Portable - 10'	Each	310.00
	Bleachers, Elevated		
	5 - Row x 15' Steel - Seats 50	Each	3,400.00
	10 - Row x 15' Steel - Seats 100	Each	5,900.00
	15 - Row x 15' Steel - Seats 150	Each	8,300.00
	Litter Receptacles		
	Pedestal - 24"	Each	310.00
	36"	Each	380.00
	Stoves	Each	370.00
	Tables - Steel Frame - Wood Slat - 8'	Each	400.00
	Aluminum Slat - 8'	Each	435.00
	All Wood	Each	255.00
	All Aluminum	Each	440.00
.4	SHELTERS	SqFt	14.00

		UNIT	COST	UNIT	COST
0209.0	**LANDSCAPING (L&M) (Laborers)**				
.1	SEEDING (.80 lb) - Machine	SqYd	.40	Acre	1,820.00
	Hand	SqYd	.90	Acre	4,450.00
	Add Fine Grading - Hand	SqYd	.80	Acre	3,860.00
	Add for 30 - Day Maintenance	-	-	Acre	1,300.00
	Deduct for Rye and Wild Grasses	-	-	Acre	580.00
.2	SODDING - Flat ($.80 per SqYd)	SqYd	2.35	-	-
	Slope (Pegged)	SqYd	2.90	-	-
	Add Fine Grading - Hand	SqYd	.90	-	-
	Add for 30 - Day Maintenance	SqYd	.40	-	-
	Remove and Haul Away Old Sod	SqYd	1.05		
.3	DISCING	SqYd	.85	Acre	4,200.00
.4	FERTILIZING	SqYd	.25	Acre	1080.00
				Ton	880.00
.5	TOP SOIL - 4" Hand - Only	SqYd	3.00	Acre	16,500.00
	Machine and Hand	SqYd	2.10	Acre	10,800.00
		CuYd	27.00	-	-
.6	TREES AND SHRUBS				
	Trees - 2"			Each	430.00
	3"			Each	590.00
	4"			Each	800.00
	5"			Each	1010.00
	6"			Each	1,500.00
	Shrubs - Small - 1' to 3'			Each	21.00
	Large - 3' to 5'			Each	42.00
.7	BEDS				
	Wood Chips, Cedar, 2" - $12 CuYd			SqYd	2.80
	Mulch, Redwood Bark - $30 CuYd			SqYd	5.80
	Stone Aggregate - $15 CuYd			SqYd	3.20
	Planting Soil, 24" - $15 CuYd			SqYd	3.60
.8	WOOD CURBS AND WALLS				
	Curbs - 4" x 4" - Fir or Pine Treated			LnFt	2.40
	Cedar			LnFt	3.10
	6" x 6" - Fir or Pine Treated			LnFt	6.15
	Cedar			LnFt	7.70
	Walls - 4" x 4" - Fir or Pine, Incl. Dead Men			SqFt	6.50
	6" x 6" - Fir or Pine, Incl. Dead Men			SqFt	9.60
	Cedar, Including Dead Men			SqFt	14.60
.9	SPRINKLER SYSTEM, WATERED AREA			CSF	34.00
0210.0	**SOIL STABILIZATION**				
.1	CHEMICAL INTRUSION			CuFt	16.00
	Pressure Grout			Gal	5.00
.2	CONCRETE INTRUSION			CuFt	29.00
.3	VIBRO FLOTATION			-	-
0211.0	**SOIL TREATMENT**				
.1	PESTICIDE			SqFt	.25
.2	VEGETATION PREVENTATIVE			SqFt	.10
0212.0	**RAILROAD WORK**				
.1	NEW WORK			LnFt	135.00
.2	REPAIR WORK			LnFt	19.00
0213.0	**MARINE WORK**				
0214.0	**TUNNEL WORK**				

0201.0 DEMOLITION

		UNIT	COST
.32	Selective Building Removals -		
	No Cutting or Disposal Included (see 0201.34 & 0201.35)		
	Concrete - Hand Work (see 0201.32 for Machine Work)		
	8" Walls - Reinforced	SqFt	10.40
	Non - Reinforced	SqFt	7.30
	12" Walls - Reinforced	SqFt	18.40
	12" Footings x 24" wide	LnFt	16.20
	x 36" wide	LnFt	22.50
	16" Footings x 24" wide	LnFt	28.30
	6" Structural Slab - Reinforced	SqFt	6.80
	8" Structural Slab - Reinforced	SqFt	8.30
	4" Slab on Ground - Reinforced	SqFt	3.15
	Non - Reinforced	SqFt	2.30
	6" Slab on Ground - Reinforced	SqFt	3.70
	Non - Reinforced	SqFt	2.80
	Stairs - Reinforced	SqFt	11.50
	Masonry - Hand Work		
	4" Brick or Stone Walls	SqFt	2.15
	4" Brick and 8" Backup Block or Tile	SqFt	3.90
	4" Block or Tile Partitions	SqFt	2.10
	6" Block or Tile Partitions	SqFt	2.20
	8" Block or Tile Partitions	SqFt	2.55
	12" Block or Tile Partitions	SqFt	3.20
	Miscellaneous - Hand Work (Including Loading)		
	Acoustical Ceilings - Attached (Including Iron)	SqFt	.92
	Suspended (Including Grid)	SqFt	.62
	Asbestos - Pipe	LnFt	71.50
	Ceilings and Walls	SqFt	23.60
	Columns and Beams	SqFt	59.00
	Tile Flooring	SqFt	2.75
	Cabinets and Tops	LnFt	16.10
	Carpet	SqFt	50.00
	Ceramic and Quarry Tile	SqFt	1.85
	Doors and Frames - Metal	Each	82.00
	Wood	Each	75.20
	Drywall Ceilings - Attached	SqFt	1.20
	Drywall on Wood or Metal Studs - 2 Sides	SqFt	1.28
	Paint Removal - Doors and Windows	SqFt	1.28
	Walls	SqFt	1.00
	Plaster Ceilings - Attached (Including Iron)	SqFt	1.70
	on Wood or Metal Studs	SqFt	1.80
	Roofing - Builtup	SqFt	1.60
	Shingles - Asphalt and Wood	SqFt	.55
	Terrazzo Flooring	SqFt	2.75
	Vinyl Flooring	SqFt	.58
	Wall Coverings	SqFt	1.05
	Windows	Each	54.00
	Wood Flooring	SqFt	.57
.33	Site Removals (Including Loading)		
	4" Concrete Walks - Labor Only (Non-Reinforced)	SqFt	2.30
	Machine Only (Non-Reinforced)	SqFt	.75
	6" Concrete Drives - Labor Only (Non-Reinforced)	SqFt	2.90
	Machine Only (Reinforced)	SqFt	1.05
	6" x 18" Concrete Curb - Machine	LnFt	1.92
	Curb and Gutter - Machine	LnFt	2.55
	2" Asphalt - Machine	SqFt	.48
	Fencing - 8' Hand	LnFt	2.75

		UNIT	COST
0202.0	**EARTHWORK**		
.1	GRADING - Hand - 4" - Site	SqFt	.25
	Hand – 4" Building	SqFt	.42
.2	EXCAVATION - Hand - Open - Soft (Sand)	CuYd	35.00
	Medium (Clay)	CuYd	44.00
	Hard (Shale)	CuYd	70.50
	Add for Trench or Pocket		15%
.3	BACK FILL - Hand - Not Compacted (Site Borrow)	CuYd	16.40
.4	BACK FILL - Hand - Compacted (Site Borrow)		
	12" Lifts - Building - No Machine	CuYd	27.80
	With Machine	CuYd	20.70
	18" Lifts - Building - No Machine	CuYd	26.10
	With Machine	CuYd	19.20
0205.0	**DRAINAGE**		
.4	BUILDING FOUNDATION DRAINAGE		
	4" Clay Pipe	LnFt	4.10
	4" Plastic Pipe - Perforated	LnFt	3.05
	6" Clay Pipe	LnFt	4.55
	6" Plastic Pipe - Perforated	LnFt	4.80
	Add for Porous Surround - 2' x 2'	LnFt	6.00
0206.0	**PAVEMENT, CURBS AND WALKS**		
.2	CURBS AND GUTTERS		
.21	Concrete - Cast in Place (Machine Placed)		
	Curb - 6" x 12"	LnFt	13.10
	6" x 18"	LnFt	15.40
	6" x 24"	LnFt	20.50
	6" x 30"	LnFt	23.10
	Curb and Gutter - 6" x 12"	LnFt	16.50
	6" x 18"	LnFt	20.10
	6" x 24"	LnFt	23.80
	Add for Hand Placed	LnFt	8.50
	Add for 2 #5 Reinf. Rods	LnFt	1.95
	Add for Curves and Radius Work	LnFt	40%
.22	Concrete Precast - 6" x 10" x 8"	LnFt	12.30
	6" x 9" x 8"	LnFt	11.15
.23	Bituminous - 6" x 8"	LnFt	3.65
.24	Granite - 6" x 16"	LnFt	38.50
.25	Timbers - Treated - 6" x 6"	LnFt	7.60
	6" x 8"	LnFt	10.50
.26	Plastic - 6" x 6"	LnFt	9.80
.3	WALKS		
	Bituminous - 1 1/2" with 4" Sand Base	SqFt	1.50
	2" with 4" Sand Base	SqFt	1.60
	Concrete - 4" - Broom Finish	SqFt	4.30
	5" - Broom Finish	SqFt	4.65
	6" - Broom Finish	SqFt	4.80
	Add for 6" x 6", 10 - 10 Mesh	SqFt	.40
	Add for 4" Sand Base	SqFt	.58
	Add for Exposed Aggregate	SqFt	1.00
	Crushed Rock - 4"	SqFt	.70
	Brick - 4" - with 2" Sand Cushion	SqFt	10.10
	4" - with 2" Mortar Setting Bed	SqFt	12.20
	Flagstone – 1 1/4" - with 4" Sand Cushion	SqFt	16.00
	1¼" - with 2" Mortar Setting Bed	SqFt	17.80
	Precast Block - 1" Colored with 4" Sand Cushion	SqFt	3.30
	2" Colored with 4" Sand Cushion	SqFt	3.80
	Wood - 2" Boards on 6" x 6" Timbers	SqFt	5.70
	2" Boards on 4" x 4" Timbers	SqFt	5.15
	Slate - 1 1/4" - with 2" Mortar Setting Bed	SqFt	21.00

* See Section 0505 for Permanent Forms of Corrugated Deck and Steel Lath.

0301.0 GENERAL CONCRETE WORK (S) (Cement Finishers & Laborers)

		UNIT	LABOR	MATERIAL
.1	CONCRETE PLACING & VIBRATING (3500# Concrete with 5.5 sack cement)			
.11	Footings - 1 1/2" Aggregate			
	Wall & Pad Types - Truck Chuted	CuYd	14.00	76.00
	Buggies	CuYd	18.00	76.00
	Crane	CuYd	19.00	76.00
.12	Wall & Grade Beams - 3/4" Aggregate			
	At Grade (to 8' deep) -Truck Chuted	CuYd	14.60	78.00
	Buggies	CuYd	19.70	78.00
	Crane	CuYd	19.00	78.00
	At Grade (over 8' deep, poured with trunks) - Truck Chuted	CuYd	15.50	78.00
	Buggies	CuYd	21.00	78.00
	Crane & Hoppers	CuYd	20.00	78.00
	Above Grade (deeper than 8') - Conveyors	CuYd	14.60	78.00
	Ramp & Buggies	CuYd	21.00	78.00
	Crane & Hoppers	CuYd	20.00	78.00
	Climbing Crane	CuYd	19.00	78.00
	Pumping	CuYd	16.00	78.00
.13	Column & Pedestals - 3/4" Aggregate - Buggies	CuYd	23.00	78.00
	Crane	CuYd	19.70	78.00
	Tower & Buggies	CuYd	23.00	78.00
	Fork Lift	CuYd	40.50	78.00
	Climbing Crane	CuYd	21.00	78.00
	Pumping	CuYd	17.00	78.00
.14	Beams - 3/4" Aggregate - Buggies	CuYd	22.00	78.00
	Crane	CuYd	20.00	78.00
	Tower & Buggies	CuYd	22.00	78.00
	Fork Lift	CuYd	40.00	78.00
	Climbing Crane	CuYd	19.60	78.00
	Pumping	CuYd	17.50	78.00
.15	Slabs - 6" Structure - 3/4" Aggregate - Buggies	CuYd	19.00	78.00
	Crane	CuYd	20.00	78.00
	Climbing Crane	CuYd	18.00	78.00
	Pumping	CuYd	17.00	78.00
.16	Stairs & Landings -			
	Structural - 3/4" Aggregate - Buggies	CuYd	22.00	78.00
	Truck Chuted	CuYd	18.00	78.00
	Climbing Crane	CuYd	20.00	78.00
	Pumping	CuYd	18.00	78.00
	Pan Filled - 1/2" Ready Mixed	CuYd	38.20	83.00
	Hand Mixed (Dry)	CuYd	59.50	75.00
.17	Curbs, Platforms & Misc. Small Pours - 1/2" Aggregate Buggies	CuYd	27.50	83.00
	Fork Lift	CuYd	38.10	83.00
.18	Slabs on Ground - 3/4" Aggregate - Truck Chuted	CuYd	15.00	78.00
	Buggies	CuYd	19.60	78.00
	Crane	CuYd	18.55	78.00
	Conveyors	CuYd	15.50	78.00
.19	Slabs-Over Metal Decks or Lath - 1/2" Aggregate Buggies	CuYd	23.00	83.00
	Tower & Buggies	CuYd	24.50	83.00
	Crane	CuYd	21.20	83.00
.20	Toppings - 1/2" Aggregate - Buggies	CuYd	22.20	83.00
	Towers & Buggies	CuYd	24.50	83.00

Conveying Equipment and Operator not included in above costs. See Page 3-5. Also other variations on Page 3-5

		UNIT	LABOR	MATERIAL
0301.0	**GENERAL CONCRETE WORK, Cont'd...**			
	Additions to Concrete Placing Work:			
	Add for Ea 500# Concrete above 3500# Concrete	CuYd	-	2.00
	Add or deduct for Sack Cement	CuYd	-	5.00
	Add for Non Corrosive	CuYd	-	8.00
	Add for High Early Cement Concrete	CuYd	-	8.00
	Deduct for 1 1/2" Aggregate	CuYd	-	1.80
	Add for 3/8" or 1/2" Aggregate	CuYd	-	3.25
	Add for Lightweight Aggregate (4000)	CuYd	2.90	16.00
	Add for Heavyweight Aggregate (Granite) (4000#)	CuYd	5.80	11.00
	Deduct for Fly Ash Concrete (3000#)	CuYd	-	2.60
	Add for Exposed Aggregate Mix	CuYd	-	9.60
	Add for Fibre Reinforcement Mix	CuYd	3.40	8.50
	Add for Less than 6 CuYd Delivery	Load	-	25.00
	Add for Hauls beyond 15 Miles	CuYd Mi	-	1.00
	Add for Time after 7 Minutes per Yard	Minute	-	90.00
	Add for Conveying Equipment and Operators:			
	Towers (Approx. 30 CuYd/ Hour)	CuYd	-	6.00
	Mobile Cranes (Approx. 25 CuYd/ Hour)	CuYd	-	7.00
	Tower Cranes (Approx. 20 CuYd/ Hour)	CuYd	-	7.50
	Fork Lifts (Approx. 5 CuYd/ Hour)	CuYd	-	7.00
	Conveyors (Approx. 50 CuYd/ Hour	CuYd	-	7.50
	Pumping (Approx. 35 CuYd/ Hour)	CuYd	-	4.50
	Add for Floors Above Grade - per Floor Placing	CuYd	1.10	-
	Add for Winter Work:			
	Productivity Loss	CuYd	4.70	-
	Heating Water and Aggregate	CuYd	-	6.00
	Calcium Chloride 1%	CuYd	-	2.20
	Insulation Blanket - Slabs (5 uses)	SqFt	.08	.13
	Walls & Beams (3 uses)	SqFt	.11	.10
	Heaters: Without Operators (Including Fuel - Floor Area)	CuYd	.42	1.60
		or SqFt	.04	.13
	Enclosures (3 uses)- Wall Area	CuYd	.90	.37
		or SqFt	.09	.16
.2	FINISHING (Including Screeds and Bulkheads to 6")			
.21	Rough Screeding - Slabs	SqFt	.32	.07
	Stairs	SqFt	.55	.07
.22	Trowel Finishing - Slabs on Ground	SqFt	.45	.07
	Solid and Pan Slabs	SqFt	.49	.07
	Slab over Corrugated	SqFt	.55	.07
	Topping Slabs	SqFt	.49	.07
	Stairs	SqFt	.83	.07
	Curbs and Bases	SqFt	.83	.07
.23	Broom Finishing - Slabs on Ground	SqFt	.42	.07
	Solid and Pan Slabs	SqFt	.48	.07
	Slab over Corrugated	SqFt	.55	.07
	Topping Slabs	SqFt	.48	.07
	Stairs	SqFt	.84	.07
.24	Float Finish - Slabs on Ground	SqFt	.39	.07
	Solid and Pan Slabs	SqFt	.43	.07
	Slab over Corrugated	SqFt	.46	.07
	Stairs	SqFt	.75	.07
.25	Special Finishing			
.251	Patch Walls, 2 Sides Tie Holes & Honeycomb	SqFt	.17	.08
.252	Rub Walls, 1 Side Carborundum for Fins	SqFt	.10	.04
	With Burlap and Grout	SqFt	.50	.08
.253	Level & Top Floors - Trowel Finish 1"	SqFt	.85	.38
	With Epoxy 1/4"	SqFt	1.65	6.00
.254	Exposed Aggregate Slabs - Washed	SqFt	.65	.06
	Retardant	SqFt	.22	.10
	Add for - Seeding	SqFt	.33	.10
	Add to All Finishing Items Above			
	Winter Production Loss and Cost	SqFt	.17	.06
	Sloped Work	SqFt	.14	05
	Heavyweight Aggregates	SqFt	.17	.06

		UNIT	LABOR	MATERIAL
0301.0	**GENERAL CONCRETE WORK, Cont'd...**			
.255	Bushhammer - Green Concrete	SqFt	1.10	.09
	Cured Concrete	SqFt	1.65	.10
	Sand Blast - Light Penetration	SqFt	.80	.10
	Heavy Penetration	SqFt	1.40	.20
.3	SPECIALTIES			
.31	Abrasives - Non-Slip			
	Alo-Grit (.86 lb & 1/4 lb/SqFt)	SqFt	.20	.24
	Carborundum Grits (1.30 lb & 1/4 lb/SqFt)	SqFt	.20	.40
	Strips 3/8" x 1/4"	LnFt	.70	1.40
	Epoxy Coating (50.00 gal) & 50 SqFt/Gal	SqFt	.30	1.50
.32	Admixtures (per CuYd Concrete)			
	Accelerator (.06 oz)	CuYd	-	1.45
	Air Entraining (.04 oz)	CuYd	-	.85
	Densifiers (.07 oz)	CuYd	-	1.80
	Retarders (.05 oz)	CuYd	-	.85
	Water Reducing (.06 oz)	CuYd	-	2.10
.33	Colors			
	Dust on Type:			
	Black, Brown & Red (.50 lb) 50#/100 SqFt	SqFt	.20	.38
	Green and Blue (.60 lb) 50#/100 SqFt	SqFt	.20	.46
	Integral (Top 1"):			
	Black, Brown & Red (1.50 lb) 27#/100 SqFt	SqFt	.38	.65
	Green and Blue (4.25 lb) 27#/100 SqFt	SqFt	.38	1.45
	Full Thickness			
	Black, Brown & Red	CuYd	23.00	120.00
	Green and Blue	CuYd	23.00	160.00
.34	Curing			
	Curing Compounds			
	Resin (10.00 gal) 200 SqFt/gal	SqFt	.08	.06
	Hydrocide Res.Base (15.00 gal) 200 SqFt/gal	SqFt	.07	.08
	Rubber Base (8.00 gal) 200 SqFt/gal	SqFt	.07	.05
	Wax Base (3.00 gal) 200 SqFt/gal	SqFt	.07	.03
	Asphalt Base (4.00 gal) 200 SqFt/gal	SqFt	.07	.03
	Paper	SqFt	.08	.07
	Polyethylene - 4 mil	SqFt	.08	.05
	Water	SqFt	.08	.01
	Burlap	SqFt	.09	.07
	Curing and Sealing	SqFt	.11	.07
.35	Expansion and Control Joints			
	Asphalt - Fibre - 1/2" x 4"	LnFt	.35	.25
	1/2" x 6"	LnFt	.36	.35
	1/2" x 8"	LnFt	.44	.48
	Polyethylene Foam - 1/2" x 4"	LnFt	.35	.42
	1/2" x 6"	LnFt	.37	.53
	1/2" x 8"	LnFt	.45	.64
	Sponge Rubber - 1/2" x 4"	LnFt	.35	2.25
	1/2" x 6"	LnFt	.36	3.30
	1/2" x 8"	LnFt	.45	4.45
	Paper - Fibre (No Oil) - 1/2" x 4"	LnFt	.35	.29
	1/2" x 6"	LnFt	.37	.39
	1/2" x 8"	LnFt	.45	.50
	Add for Cap	LnFt	.20	.30
	Add for 3/4" Thickness	LnFt	-	30%
	Add for 1" Thickness	LnFt	-	80%

See 0413.5 for Other Expansion Joint Costs.
See 0410.0 Mortars for Cement Prices.

		UNIT	LABOR	MATERIAL
0301.0	**GENERAL CONCRETE WORK, Cont'd...**			
.36	Grouts - Non-Shrink			
	Iron Oxide (.45/lb)			
	Hand Mixed (1 Cem: 1 Sand: 1 Iron Oxide)	CuFt	6.40	27.00
	Per 1"	SqFt	.80	2.20
	Premixed (.45/lb)	CuFt	2.30	39.00
	Aluminum Oxide (.45/lb)	CuFt	4.60	12.00
	Non-Metallic, Premixed (.35/lb)	CuFt	3.40	29.00
.37	Hardeners and Sealers			
	Acrylic Sealer (7.50 gal) 300 SqFt/gal	SqFt	.12	.05
	Epoxy Sealer (25.00 gal) 300 SqFt/gal	SqFt	.12	.13
	Urethane Sealer (9.50 gal) 400 SqFt/gal	SqFt	.12	.05
	Liquid Hardeners (4.85 gal) 200 SqFt/gal	SqFt	.12	.05
.38	Joint Sealers			
	Rubber Asphalt			
	Hot 1/2" x 1/2" Joint (.70 lb)	LnFt	.57	.50
	Cold 1/2" x 1/2" Joint (.54 lb)	LnFt	.40	.57
	Epoxy (20.00 Qt)	LnFt	.58	1.45
.39	Moisture Proofing (Loose Laid for Slabs)			
	Polyethylene 4 Mil	SqFt	.07	.03
	6 Mil	SqFt	.07	.04
	Asphalted Paper	SqFt	.07	.05
.4	EQUIPMENT (Conv., Finishing, Vibrating, etc.)			
	See Divisions I-6A and I-7A			
.5	TESTS			
	Cylinder - 6" x 12" (7-day and 28-day)	Ea	-	40.00
	Add per Pickup	Ea	-	40.00
0302.0	**CONCRETE FORM WORK (S) (Carpenters)**			
	(Lumber @ 500.00 MBF & 3/4" BB Plywood @ 1.05 SqFt)			
	(Costs Include Erection, Stripping, Cleaning, Oiling)			
.1	REMOVABLE FORMS (Expendable and Reusable)			
.11	Footings (2-1/2 BdFt/SqFt and 3 Uses)			
	Wall Type - Constant Elevation	SqFt	2.20	.59
	Pad Type	SqFt	2.30	.59
	Add for Volume Elevations and Changes	SqFt	.16	-
	Add for Deep Foundation (8' or more)	SqFt	.16	-
	Add for Each Foot Form Deeper than 12"	SqFt	.10	.05
	Add for Keyway	LnFt	.45	.12
.12	Walls & Grade Beams (2-1/2 BdFt/SqFt & 3 Uses)			
	Straight Walls - 4' High	SqFt	2.25	.70
	Add for Each Foot Higher than 4'	SqFt	.05	.04
	Add for Walls Above Grade	SqFt	.06	-
	Add for Deep Foundation (8' or more)	SqFt	.12	-
	Add for Pilastered Wall, 24' O.C. Total Area	SqFt	.11	.03
	Add for Pilastered Wall, Pilasters Only	SqFt	4.40	.60
	Add for Openings - Opening Area	SqFt	2.90	.75
	Add for Retaining and Battered Type Wall	SqFt	.86	.12
	Add for Curved Wall	SqFt	1.75	.15
	Add for Brick Ledge	LnFt	1.70	.16
	Add for Parapet Wall - Hung	SqFt	2.65	.27
	Add for Pit or Small Trench Walls	SqFt	1.58	.10
	Add for Lined Forms - Hardwood Panels	SqFt	.38	.48
	Gang Formed Walls - Make Up	SqFt	4.00	6.00
	Move	SqFt	.92	.20
	3/4" B-B Plywood - Avg - 5 Ply - 8 Uses @ 1.05	SqFt	-	-
	7 Ply - 8 Uses @ 1.20	SqFt	-	-
	M.D.O. Plywood - Avg. - 20 Uses @ 1.25	SqFt	-	-

0302.0 CONCRETE FORM WORK, Cont'd...

		UNIT	LABOR	MATERIAL
.13	Columns & Pedestals			
	Sq. & Rectangular (2 1/2 BdFt/ SqFt & 3 Uses)			
	8" x 8"	SqFt	4.40	.83
	12" x 12"	SqFt	4.35	.84
	16" x 16"	SqFt	4.30	.85
	20" x 20"	SqFt	4.15	.86
	24" x 24"	SqFt	4.20	.87
	Add per Foot - Work over 10' Floor Heights	SqFt	.35	-
	Add per Floor above 20' Above Grade	SqFt	.14	-
	Deduct for Ganged Formed (8 Uses)	SqFt	1.20	-
	Round - Steel or Fiberglass (Rent and 3 Uses)			
	12"	LnFt	6.75	4.05
	16"	LnFt	6.55	4.50
	20"	LnFt	7.70	5.30
	24"	LnFt	11.30	6.20
	30"	LnFt	14.75	7.50
	36"	LnFt	18.35	9.30
	Add per Foot - Work over 10" Floor Heights	LnFt	.35	.40
	Round-Fibre (6" to 48" x 18' available)			
	8" Cut Length Prices Used	LnFt	4.70	2.25
	12"	LnFt	4.90	4.30
	16"	LnFt	4.30	6.90
	20"	LnFt	6.70	9.50
	24"	LnFt	9.10	12.50
	30"	LnFt	10.60	15.50
	36"	LnFt	13.10	17.60
	Add for Less than 100 Feet - One Size	LnFt	-	20%
	Add for Seamless Fibre	LnFt	-	15%
	Add for Conical Heads to Steel or Fibre	Each	93.00	57.00
	Add for Beam Fittings or Other Openings	Each	104.00	62.00
.14	Beams (12" Floor Heights and 3 Uses Lumber)			
	Spandrel Beams (Includes Shoring)			
	12" x 48" (4 1/2 BdFt/ SqFt)	SqFt	4.80	1.08
	12" x 42" (4 3/4 BdFt/ SqFt)	SqFt	4.85	1.10
	12" x 36" (5 BdFt/ SqFt)	SqFt	4.95	1.13
	12" x 30" (5 1/4 BdFt/ SqFt)	SqFt	5.05	1.16
	8" x 42" (4 3/4 BdFt/ SqFt)	SqFt	5.05	1.10
	8" x 36" (5 1/4 BdFt/ SqFt)	SqFt	5.25	1.15
	Add for Decks and Safety Rail	SqFt	1.00	.35
	Interior Beams			
	16" x 30" (4 BdFt/ SqFt)	SqFt	4.80	1.07
	16" x 24" (4 1/2 BdFt/ SqFt)	SqFt	4.85	1.10
	12" x 30" (4 3/4 BdFt/ SqFt)	SqFt	4.90	1.12
	12" x 24" (5 1/4 BdFt/ SqFt)	SqFt	4.85	1.18
	12" x 16" (6 1/2 BdFt/ SqFt)	SqFt	5.35	1.30
	8" x 24" (5 1/4 BdFt/ SqFt)	SqFt	5.20	1.20
	8" x 16" (6 1/2 BdFt/ SqFt)	SqFt	5.45	1.30
	Add for Beams Carrying Horizontal Shoring	SqFt	.92	.30
	Add per Foot for Floor Heights over 10'	SqFt	.45	.10
	Add per Floor over 20' Above Grade	SqFt	.14	.09
	Add for Splayed Beams	SqFt	1.62	.18
	Add for Inverted Beams	SqFt	2.04	.19
	Add for Mud Sills (3 Uses)	SqFt	.76	.35

0302.0	CONCRETE FORM WORK, Cont'd...	UNIT	LABOR	MATL
.15	Slabs (12' Floor Heights and 3 Uses Lumber)			
.151	Solid Slabs (Includes Shoring) (Metal Adj. Beams or Joists)			
	Horizontal Shoring Method			
	To 5" thick (2' OC x 10' Span) 2.5 BdFt/ SqFt	SqFt	1.90	.65
	5" - 8" thick (2' OC x 15' Span) 2.8 BdFt/ SqFt	SqFt	1.95	.72
	8" - 11" thick (2' OC x 20' Span) 3.1 BdFt/ SqFt	SqFt	2.20	.78
	Add for Heights above 10' per foot	SqFt	.25	.08
	Vertical Shoring Method			
	(Wood or Metal Posts, Purlins & Joists)			
	To 5" thick - 2.7 BdFt/SqFt - 4' x 5'6" OC	SqFt	1.95	.70
	5" - 8" thick - 3.0 BdFt/SqFt - 4' x 5'0" OC	SqFt	2.15	.76
	8" - 11" thick - 3.3 BdFt/SqFt - 4' x 4'6" OC	SqFt	3.65	.80
	Add for Heights above 10' per foot	SqFt	.24	.08
	Add for Adjustable Hardware	SqFt	-	.02
	Add for Cantilevered Slabs	SqFt	1.30	.15
	Add for Drop Panels - No Edge included	SqFt	.24	.10
	Flying Form method (8 Uses)			
	To 5" Thick	SqFt	1.62	.60
	5" - 8" Thick	SqFt	1.67	.61
	8" - 11" Thick	SqFt	1.77	.62
	Add for Heights above 8' per foot	SqFt	.22	.07
	Add for 2 Uses	SqFt	.20	.25
	Add for 4 Uses	SqFt	.15	.18
.152	Pan Slabs - Cost same as solid slabs above plus pan forming costs below.			
	Based on 3 Uses and lease of approximately 4,000 Square Feet			
	20" Pan 8" + 2"	SqFt	.90	.72
	10" + 2"	SqFt	.92	.74
	12" + 2"	SqFt	.95	.76
	14" + 2"	SqFt	1.00	.79
	30" Pan 8" + 2 1/2"	SqFt	.87	.69
	10"+2 1/2"	SqFt	.89	.71
	12"+2 1/2"	SqFt	.92	.74
	14"+2 1/2"	SqFt	.95	.76
	19" x 19" Dome Pan - 24" x 24" Joist Center			
	6" + 2"	SqFt	.86	.71
	8" + 2"	SqFt	.89	.73
	10" + 2"	SqFt	.91	.74
	12" + 2"	SqFt	.96	.77
	30" x 30" Dome Pan - 36" x 36" Joist Center			
	10" + 2 1/2"	SqFt	.96	.78
	12" + 2 1/2"	SqFt	1.03	.80
	14" + 2 1/2"	SqFt	1.10	.84
	Add for 2 Uses	SqFt	.06	.10
	Add for 1 Use	SqFt	.07	.23
	Deduct for 4 Uses	SqFt	.06	.07
	Deduct for 5 Uses	SqFt	.05	.08
	Pan Slabs: Based on 5 Uses & Lease of 4,000 SqFt			
	40" x 40" Dome Pan - 16" + 3"	SqFt	1.20	1.05
	48" x 48" Joist Center - 18" + 3"	SqFt	1.30	1.25
	Deduct for 6 Uses	SqFt	.06	.10
	Deduct for 7 Uses	SqFt	.05	.11
	Add to above units if subcontracted		30%	10%

		UNIT	LABOR	MATL
0302.0	**CONCRETE FORM WORK, Cont'd...**			
.16	Stairs and Landing (3 BdFt and 2 Uses)			
	Structural - Contact Area of			
	Soffits, Stringers and Risers	SqFt	4.10	1.20
	On Ground - Contact Area of			
	Stringers and Risers	SqFt	5.35	1.25
.17	Curbs & Platforms (2 BdFt and 2 Uses)	SqFt	3.80	.72
.18	Edge Forms/Bulkheads for Slabs (2 BdFt, 2 Uses)			
	Floor Supported	SqFt	3.70	.72
	Structural	SqFt	5.10	.80
.2	SHORING, SUPPORTS AND ACCESSORIES (All items in this section are included in SqFt costs of Section 0302.1. This section is for cost-keeping & limited analyzing of variables.)			
.21	Horizontal Shoring			
	10' Span - Monthly Rent Cost	Each		1.80
	SqFt Contact Area	SqFt		.40
	15' Span - Monthly Rent Cost	Each		6.75
	SqFt Contact Area	SqFt		.50
	20' Span - Monthly Rent Cost	Each		7.90
	SqFt Contact Area	SqFt		.68
.22	Vertical Shoring (4' OC - 10' High) - Average	SqFt		.30
	Wood Posts with Ellis Shore Hardware			
	Single - W/T Head - New	Each		12.50
	LnFt Beam - 6 Uses	LnFt		.64
	Double - W/T Head & Bracing - New	Each		26.00
	LnFt Beam - 6 Uses	LnFt		1.15
	Metal Posts - Single - Monthly Rent Cost	Each		3.65
	LnFt Beam - 2 Uses per Month	LnFt		.82
	Double - W/T Head & Bracing	LnFt		6.75
	LnFt Beam - 2 Uses per Month	LnFt		1.55
	Tubular Frames - to 10'			
	Mo. Cost - Incl. Head & Base - 10,000#	Each		14.60
	LnFt - Average - 2 Uses per Month	LnFt		1.25
	Deduct for 6,000#	Each		25%
.23	Column Clamps and Steel Strapping			
	Clamps - Monthly Rent Cost - Steel	Each		4.45
	Lock Fast	Each		8.30
	Gang Form	Each		7.25
	Contact Area - 2 Uses per Month	SqFt.		37
	Strapping - 3/4" x .025 (1.10 lb)	LnFt		.28
	3/4" x .023 (1.00 lb)	LnFt		.25
	Contact Area of Forms	SqFt		.15
.24	Wall and Beam Brackets			
	Brackets (2' OC) - Monthly Rent Cost	Each		.62
	New Cost	Each		6.00
	Contact Area - 4 Uses per Month	SqFt		.15
.25	Accessories - Beam Hangers			
	Tyloops, Lags, Washers & Bolts - Heavy	LnFt		.90
	Light	LnFt		.65

0302.0 CONCRETE FORM WORK, Cont'd...

		UNIT	LABOR	MATERIAL
.3	SPECIALTIES			
.31	Nails, Wire and Ties			
	Nails 6 d common (50# box)	CWT	-	51.00
	8 d common	CWT	-	52.00
	16 d common	CWT	-	56.00
	8 d box	CWT	-	80.00
	16 d box	CWT	-	84.00
	9 ga concrete nails (3/4" to 3")	CWT	-	140.00
	Add for Coated Nails	CWT	-	60%
	Wire - Black Annealed 9 ga (100# Coil)	CWT	-	80.00
	16 ga (100# Coil)	CWT	-	77.00
	Average Nail and Wire Cost for Formwork	BdFt	-	.02
		or SqFt	-	.03
	Ties - Breakback -8" 3000# 5" ends	Each	-	.85
	12" 3000# 5" ends	Each	-	1.00
	16" 3000# 5" ends	Each	-	1.05
	Add for 5000#	Each	-	.14
	Add for 8" ends	Each	-	.08
	Average Tie Cost for Formwork (2' OC)			
	Wall Area	SqFt	-	.36
	Contact Area Form	SqFt	-	.20
	Gang Form Contact Area	SqFt	-	.21
	Coil Bolts - 8"	Each	-	1.30
	12"	Each	-	1.45
.32	Anchors and Inserts (including Layout)			
	Anchor Bolts (Concrete) -			
	1/2" x 12"	Each	1.05	.75
	5/8" x 12"	Each	1.15	1.80
	3/4" x 12"	Each	1.30	2.70
	Ceiling Type Inserts			
	Adjustable - 1/2" Bolt Size x 3'	Each	.85	2.20
	5/8" Bolt Size x 3'	Each	.90	2.25
	3/4" Bolt Size x 3'	Each	.95	2.70
	Continuous Slotted	Each	.75	1.70
	Threaded - 1/2" Bolt Size	Each	.88	3.75
	5/8" Bolt Size	Each	.90	6.20
	3/4" Bolt Size	Each	1.00	7.30
	Hanger Wire Type - Drive In	Each	.75	.18
	Shell	Each	.75	.22
	Ferrule Type - 1/2"	Each	.80	1.05
	5/8"	Each	.85	1.25
	3/4"	Each	.90	1.75
	T-Hanger - 4"	Each	.85	.95
	10"	Each	.85	1.40
	14"	Each	1.00	1.90
	Wall and Beam Type Insert			
	Dovetail Slot - 24 ga (No Anchors)	LnFt	.32	.40
	22 ga (No Anchors)	LnFt	.41	.80
	Flashing Reglects	LnFt	.55	.65
	Shelf Angle Inserts - Add for Bolts,			
	Washers, and Nuts -5/8"	Each	.85	2.85
	3/4"	Each	.90	3.60

0302.0	CONCRETE FORM WORK, Cont'd...	UNIT	LABOR	MATERIAL
	Bolts - 5/8" x 2"	Each	.77	1.10
	3/4" x 2"	Each	.78	1.70
	3/4" x 3"	Each	.80	2.00
	Nuts - 5/8"	Each	.28	.25
	3/4"	Each	.32	.35
	Washers	Each	.07	.25
	Threaded Rod - 3/8" x 36" (Galv)	Each	1.00	1.30
	1/2" x 36" (Galv)	Each	1.15	2.40
	5/8" x 36" (Galv)	Each	1.35	3.50
.33	Chamfers, Drips and Reveals (1 Use)			
	Chamfers - 3/4" Metal	LnFt	.26	.48
	Plastic with Tail	LnFt	.28	.52
	without Tail	LnFt	.30	.43
	Wood	LnFt	.26	.15
	Reveals - 3/4" Plastic	LnFt	.38	1.25
	Drips	LnFt	.34	.42
.34	Stair Nosings and Treads			
	Nosings, Curb Bar - Steel - Galvanized	LnFt	1.10	9.00
	Treads 3" x 1/2" - Cast Iron	LnFt	1.35	10.50
	Extruded Aluminum	LnFt	1.35	7.00
	Cast Aluminum	LnFt	1.35	9.00
.35	Water Stops			
	Polyvinyl Chloride - 6" x 3/16"	LnFt	.80	1.35
	6" Centerbulb	LnFt	.84	2.55
	9" x 3/16"	LnFt	.90	2.00
	9" Centerbulb	LnFt	1.00	3.25
	Add for 3/8"	LnFt	.17	50%
	Rubber (Neoprene) - 6" Centerbulb	LnFt	.88	8.20
	9" Centerbulb	LnFt	1.00	16.00
	Add per Splice	Each	8.00	-
.36	Divider Strips - White Metal, 16 ga x 1 1/4"	LnFt	.67	.90
	1/2" x 1 1/4"	LnFt	.68	.95
	3/16" x 1 1/4"	LnFt	.67	.95
	1/4" x 1 1/4"	LnFt	.67	1.60
	Add for Brass 1/8"	LnFt	.12	.85
.37	Tongue and Groove Joint Forms			
	Asphalt - 3 1/2" x 1/8"	LnFt	.73	.75
	5 1/2" x 1/8"	LnFt	.80	.88
	Metal - 3 1/2"	LnFt	.74	.62
	5 1/2"	LnFt	.80	.80
	Wood - 3 1/2"	LnFt	.74	.24
	5 1/2"	LnFt	.80	.38
	Plastic - 3 1/2"	LnFt	.74	.85
	Add for Stakes	Each	.72	.54
.38	Bearing Pads and Shims			
	Vinyl - 1/8" x 6" x 24"	SqFt	.82	2.00
	1/4" x 6" x 24"	SqFt	.95	4.10
	Neoprene - 1/8" x 36" (70 Duro)	SqFt	.70	5.10
	1/4" x 36"	SqFt	.75	9.25
.4	EQUIPMENT			

See Pages 1 - 6A and 1 - 7A. Average 5% Labor Cost.

Add to Labor Formwork Costs Above for Winter Construction 10% to 25%

0303.0	REINFORCING STEEL (S&L) (Ironworkers)	UNIT	LABOR	MATERIAL	
			Subcont	Plain	Epoxy
.1	BARS (Based on Local Trucking)				
	1/4" to 5/8" Large Size Job (100 Tons Up)	Ton	460	1010	1,800
	Medium Job (20 - 99 Tons Up)	Ton	515	1050	1,860
	Small Size Job (5 - 19 Tons)	Ton	540	1100	1,970
	3/4" to 1 1/4" Large Job (100 Tons Up)	Ton	475	1010	1,780
	Medium Job (10 - 99 Tons)	Ton	485	1020	1,800
	Small Size Job (5 - 19 Tons)	Ton	510	1060	1,880
	1 1/2" to 2" Average	Ton	500	1040	1,840
	Add For Galvanizing	Ton	60	460	-
	Add for Bending - Light #2 to #5	Ton	-	250	-
	Heavy #6 and Larger	Ton	-	130	-
	Add for Size Extras - Light #2 to #4	Ton	-	150	-
	Heavy #3 and Larger	Ton	-	120	-
	Add for Spirals - Shop Assembled - Hot Rolled	Ton	-	170	-
	Cold Rolled	Ton	-	240	-
	Unassembled - Hot Rolled	Ton	-	145	-
	Cold Rolled	Ton	-	225	-
	Add for Splicing: #11 Bars - Buttweld	Each	33.00	3.00	-
	#14 Bars - Buttweld	Each	44.00	4.50	-
	#11 Bars - Mech. Buttweld	Each	46.00	11.50	-
	#14 Bars - Mech. Buttweld	Each	52.00	12.50	-

					UNIT	LABOR	MATERIAL
	BARS (Based on Local Trucking)						
	Tables of Weights & Prices by LnFt -				Ton	-	1100.00
	Small job from warehouse stock for local delivery				Lb	-	.40
	# 2 Bar	1/4" or	.250	.167 lb	LnFt	.14	.14
	# 3 Bar	3/8" or	.375	.376 lb	LnFt	.17	.21
	# 4 Bar	1/2" or	.500	.668 lb	LnFt	.23	.27
	# 5 Bar	5/8" or	.625	1.043 lb	LnFt	.33	.34
	# 6 Bar	3/4" or	.750	1.502 lb	LnFt	.44	.43
	# 7 Bar	7/8" or	.875	2.044 lb			
	# 8 Bar	1" or	1.000	2.670 lb			
	# 9 Bar	1 1/8" or	1.128	3.400 lb			
	#10 Bar	1 1/4" or	1.270	4.303 lb			
	#11 Bar	1 3/8" or	1.410	5.313 lb			
	#14 Bar	1 1/2" or	1.693	7.650 lb			
	#18 Bar	2 1/4" or	2.257	13.600 lb			

		UNIT	LABOR	MATERIAL
.2	ACCESSORIES			
	Additional sizes, but below are close to standard and are included in tonnage costs above			
	1/4" to 1 1/2" Slab Bolsters	LnFt	-	.32
	1/2" to 2" Beam Bolsters - Upper & Lower	LnFt	-	.60
	2" to 3" Beam Bolsters - Upper & Lower	LnFt	-	.95
	3" High Chairs - Continuous	LnFt	-	.50
	3/4" x 4" to 6" Joist Chairs	Each	-	.45
	1/2" Dowel Bar Tubes - Plastic and Metal	Each	.48	.30
	Add for Galvanized Accessories	-	-	.17
	Add for Plastic Accessories	-	-	.20
.3	POST TENSION STEEL - PRESTRESSED IN FIELD			
	Ungrouted - 100 kips	Lb	.78	1.75
	200 kips	Lb	.72	1.80
	Deduct for Added 100 kips	Lb	.02	-
	Add for Chairs	Lb	-	.13

		UNIT	LABOR	MATERIAL
0303.0	**REINFORCING STEEL, Cont'd...**			
.4	WIRE FABRIC (750 SqFt Roll or 120 SqFt Sheet)			
	6" x 6" x 10/10 (W 1.4)	SqFt	.14	.13
	6" x 6" x 8/8 (W 2.1)	SqFt	.15	.18
	6" x 6" x 6/6 (W 2.9)	SqFt	.19	.23
	6" x 6" x 4/4 (W 4.0)	SqFt	.23	.26
	4" x 4" x 4/4 (W 4.0)	SqFt	.28	.37
	4" x 6" x 8/8 (W 2.0)	SqFt	.19	.20
	Add for Epoxy Coated (120 SqFt Sheet)	SqFt	.02	.22

		UNIT		
0304.0	**SPECIALTY PLACED CONCRETE (No Fees)**			
.1	LIFT SLAB CONSTRUCTION			
	Concrete - 6" Slab	SqFt		3.95
	Formwork	SqFt		1.20
	Reinforcing Steel	SqFt		1.35
	Lifting	SqFt		1.35
	Brace and Assemble	SqFt		.85
			TOTAL	8.70
.2	POST TENSIONED OR STRESSED IN PLACE CONCRETE			
	Concrete - 6" Slab and Columns	SqFt		4.60
	Formwork (Assume Flying Forms)	SqFt		4.90
	Post Tensioning and Reinforcing Steel	SqFt		3.25
			TOTAL	12.75
.3	TILT UP CONSTRUCTION (Incl. Concrete Columns)			
	Concrete - 6" Slab	SqFt		4.30
	Formwork	SqFt		2.85
	Reinforcing Steel	SqFt		2.75
	Tilting Up and Bracing	SqFt		1.30
			TOTAL	11.20

		UNIT	1"	2"	3"
.4	PNEUMATICALLY PLACED CONCRETE (Gunite)				
	Walls - 1" Layers with Mesh	SqFt	3.20	4.95	6.10
	Columns - 1" Layers with Mesh	SqFt	6.10	6.60	7.70
	Roofs	SqFt	4.00	4.65	5.40

		Spans to Approximate	UNIT	COST
0305.0	**PRECAST PRESTRESSED CONCRETE (L&M)**			
.1	BEAMS AND COLUMNS			
	Beams	20'	LnFt	95.00
		30'	LnFt	112.00
	Columns - 12" x 12" x 12' with Plates		LnFt	120.00
	16" x 16" x 12' with Plates		LnFt	148.00
.2	DECKS (Flat) - 24", 40", 48", 72" and 96" Wide - 50 lb. Roof Load			
	4"	12'	SqFt	5.00
	6"	18'	SqFt	5.60
	8"	24'	SqFt	6.00
	10"	30'	SqFt	6.70
	12"	50'	SqFt	8.10
.3	DOUBLE T (Flat)			
	8' x 20" - 30" Deep - 60 lb. Roof Load	40'	SqFt	6.50
		50'	SqFt	6.85
		60'	SqFt	8.00
		70'	SqFt	8.80
		80'	SqFt	9.50
	SINGLE T			
	8' x 36" - 48" Deep	60'	SqFt	10.50
		80'	SqFt	11.35
		100'	SqFt	12.80
		120'	SqFt	14.30
	Add for Floor Decks (100# Load)		SqFt	10%
	Add for Sloped Installation to Decks & T's		SqFt	10%
	Add for 2" Topping for Floors - If Required		SqFt	3.35
	Add per Mobilization		Each	3,000.00

			UNIT		COST
0305.0	**PRECAST PRESTRESSED CONCRETE, Cont'd...**				
.4	WALLS				
	Add to Flat Decks 0305.2 Above		SqFt		1.80
	Add to Double T 0305.3 Above		SqFt		1.95
	Add for Core Wall Type		SqFt		1.75
	Add for Insulated Type		SqFt		1.85
0306.0	**PRECAST CONCRETE (M)**				
.1	BEAMS AND COLUMNS- Same as 0305.1				
.2	DECKS- Same as 0305.2				
.3	PANELS OR FACING UNITS (L&M)				
	Insulated - 8' x 16' x 12"				
	Exposed Aggregate				
	Gravel		SqFt		14.10
	Quartz & Marble		SqFt		16.20
	Trowel or Brush				
	Grey Cement		SqFt		13.70
	White Cement		SqFt		14.90
	Deduct for Non-Insulated 8' x 16' x 8"		SqFt		1.25
	Add for Banded		SqFt		.50
.4	PLANK (S)				
	2" x 24" 6' Span		SqFt		4.65
	3" x 24" 8' Span		SqFt		5.20
				LABOR	MATERIAL
.5	SPECIALTIES (S)				
	Curbs 6" x 10" x 8'		LnFt	2.20	6.70
	8" x 10" x 8'		LnFt	2.70	7.30
	Copings 5" x 13"		LnFt	6.00	9.70
	Sills & Stools 5" x 6"		LnFt	4.80	12.00
	Splash Blocks 3" x 36" x 16"		Each	9.75	24.00
	5" x 36" x 16"		Each	12.30	25.40
	8" x 36" x 16"		Each	16.20	29.70
	Medians		LnFt	9.10	38.90
0307.0	**PRECAST CEMENTITIOUS PLANK (L&M)**				
.1	EXPANDED MINERALS (Includes Clips, Rods & Grout)				COST
		SPANS TO APPROX.			
	3" x 24"	6'-0"	SqFt		3.60
	4" x 24"	9'-6"	SqFt		3.90
.2	GYPSUM (Metal Edged)				
	2" x 15"	4'-0"	SqFt		3.70
	2" x 15"	7'-0"	SqFt		4.00
.3	WOOD FIBRE				
	2" x 32"	3'-0"	SqFt		3.45
	2 1/2" x 32"	3'-6"	SqFt		3.60
	3" x 32"	4'-0"	SqFt		4.00
0308.0	**POURED IN PLACE CEMENTITIOUS CONCRETE (L&M)**				
.1	EXPANDED MINERALS - 2"		SqFt		2.25
.2	EXPANDED MINERALS & ASPHALT - 2"		SqFt		2.50
.3	GYPCRETE - 1/4"		SqFt		1.15
	1"		SqFt		1.45

All examples below are representative samples and have great variables because of design efficiencies, size, spans, weather, etc. All figures include a contractor's fee of 10% for Footings and Slabs on Ground (low risk) and a 15% fee for Walls, Columns, Beams and Slabs (high risk). All totals include material prices as follows: Concrete $78.00 CuYd, Lumber $500.00 M.B.F. (3 uses), Plywood $1.05 SqFt, and Reinforcing Steel $1050.00 Ton. Taxes and insurance on Labor added at 35%. Includes General Conditions at 10%. CuYd Costs are for information only--not quick estimating.

	Unit	Unit Cost With ReinSteel	Unit Cost Without ReinSteel	CuYd Cost With ReinSteel
FOOTINGS (Including Hand Excavation)				
Continuous 24" x 12" (3 # 4 Rods)	LnFt	20.10	18.30	271.00
36" x 12" (4 # 4 Rods)	LnFt	25.80	26.00	232.00
20" x 10" (2 # 5 Rods)	LnFt	17.10	16.50	330.00
16" x 8" (2 # 4 Rods)	LnFt	14.55	17.10	436.00
Pad 24" x 24" x 12" (4 # 5 E.W.)	Each	71.00	53.00	480.00
36" x 36" x 14" (6 # 5 E.W.)	Each	142.00	101.00	426.00
48" x 48" x 16" (8 # 5 E.W.)	Each	235.00	196.00	400.00
WALLS (#5 Rods 12" O.C. - 1 Face)				
8" Wall (# 5 12" O.C. E.W.)	SqFt	14.00	11.60	560.00
12" Wall (# 5 12" O.C. E.W.)	SqFt	15.30	14.00	413.00
16" Wall (# 5 12" O.C. E.W.)	SqFt	17.10	15.80	342.00
Add for Steel - 2 Faces - 12" Wall	SqFt	1.85	-	50.00
Add for Pilastered Wall - 24" O.C.	SqFt	.85	-	22.50
Add for Retaining or Battered Type	SqFt	2.55	-	70.00
Add for Curved Walls	SqFt	4.70	-	137.00
COLUMNS				
Sq Cornered 8" x 8" (4 # 8 Rods)	LnFt	34.20	24.70	2,050.00
12" x 12" (6 # 8 Rods)	LnFt	56.50	45.80	1,525.00
16" x 16" (6 # 10 Rods)	LnFt	77.10	58.00	1,155.00
20" x 20" (8 # 20 Rods)	LnFt	106.50	74.40	1,030.00
24" x 24" (10 # 11 Rods)	LnFt	131.00	89.00	885.00
Round 8" (4 # 8 Rods)	LnFt	24.40	17.00	1,725.00
12" (6 # 8 Rods)	LnFt	37.90	23.80	1,230.00
16" (6 # 10 Rods)	LnFt	52.00	33.50	1,040.00
20" (8 # 20 Rods)	LnFt	73.20	49.30	945.00
24" (10 # 11 Rods)	LnFt	108.00	68.80	880.00
BEAMS				
Spandrel 12" x 48" (33 # Rein. Steel)	LnFt	132.00	104.00	890.00
12" x 42" (26 # Rein. Steel)	LnFt	115.50	90.50	886.00
12" x 36" (21 # Rein. Steel)	LnFt	93.60	79.00	842.00
12" x 30" (15 # Rein. Steel)	LnFt	85.00	78.25	918.00
8" x 48" (26 # Rein. Steel)	LnFt	109.00	103.00	1,070.00
8" x 42" (21 # Rein. Steel)	LnFt	104.50	93.50	1,207.00
8" x 36" (16 # Rein. Steel)	LnFt	96.00	80.70	1,295.00
Interior 16" x 30" (24 # Rein. Steel)	LnFt	76.30	70.10	610.00
16" x 24" (20 # Rein. Steel)	LnFt	75.20	65.80	750.00
12" x 30" (17 # Rein. Steel)	LnFt	82.20	66.20	880.00
12" x 24" (14 # Rein. Steel)	LnFt	62.80	55.30	847.00
12" x 16" (12 # Rein. Steel)	LnFt	52.80	43.20	1,050.00
8" x 24" (13 # Rein. Steel)	LnFt	62.00	52.70	1,240.00
8" x 16" (10 # Rein. Steel)	LnFt	47.50	41.40	1,425.00

SLABS	Unit	Unit Cost With ReinSteel	Unit Cost Without ReinSteel	CuYd Cost With ReinSteel
Solid -				
4" Thick	SqFt	9.00	7.25	730.00
5" Thick	SqFt	10.40	8.00	675.00
6" Thick	SqFt	11.40	7.60	615.00
7" Thick	SqFt	12.90	10.50	575.00
8" Thick	SqFt	14.50	12.35	580.00
Deduct for Post Tensioned Slabs	SqFt	.80	-	-
Pan -				
Joist -20" Pan - 10" x 2"	SqFt	12.50	11.15	-
12" x 2"	SqFt	14.00	11.30	-
30" Pan - 10" x 2 1/2"	SqFt	12.50	10.70	-
12" x 2 1/2"	SqFt	14.00	12.00	-
Dome -19" x 19" - 10" x 2"	SqFt	13.50	10.80	-
12" x 2"	SqFt	14.20	10.90	-
30" x 30" - 10" x 2 1/2"	SqFt	12.30	10.60	-
12" x 2 1/2"	SqFt	13.35	10.70	-
COMBINED COLUMNS, BEAMS AND SLABS				
20' Span	SqFt	18.35	-	-
30' Span	SqFt	19.40	-	-
40' Span	SqFt	20.70	-	-
50' Span	SqFt	22.30	-	-
60' Span	SqFt	26.60	-	-
STAIRS (Including Landing)				
4' Wide 10' Floor Heights 16 Risers	Riser	161.00	-	1,330.00
5' Wide 10' Floor Heights 16 Risers	Riser	195.00	-	1,410.00
6' Wide 10' Floor Heights 16 Risers	Riser	220.00	-	1,360.00

SLABS ON GROUND	Unit	Unit Cost With Mesh	CuYd Cost With Mesh
4" Concrete Slab (6 6/10 - 10 Mesh - 5 1/2 Sack Concrete, Trowel Finished, Cured & Truck Chuted)	SqFt	3.40	275.00
Add per Inch of Concrete	SqFt	.55	
Add per Sack of Cement	SqFt	.14	
Deduct for Float Finish	SqFt	.08	
Deduct for Brush or Broom Finish	SqFt	.05	
Add for Runway and Buggied Concrete	SqFt	.17	
Add for Vapor Barrier (4 mil)	SqFt	.14	
Add for Sub - Floor Fill (4" sand/gravel)	SqFt	.40	
Add for Change to 6 6/8 - 8 Mesh	SqFt	.08	
Add for Change to 6 6/6 - 6 Mesh	SqFt	.14	
Add for Sloped Slab	SqFt	.20	
Add for Edge Strip (sidewalk area)	SqFt	.22	
Add for ½" Expansion Joint (20' O.C.)	SqFt	.07	
Add for Control Joints (keyed/ dep.)	SqFt	.16	
Add for Control Joints (joint filled)	SqFt	.17	
Add for Control Joints - Saw Cut (20' O.C.)	SqFt	.30	
Add for Floor Hardener (1 coat)	SqFt	.13	
Add for Exposed Aggregate - Washed Added	SqFt	.40	
Retarding Added	SqFt	.36	
Seeding Added	SqFt	.40	
Add for Light Weight Aggregates	SqFt	.40	
Add for Heavy Weight Aggregates	SqFt	.28	
Add for Winter Production Loss/Cost	SqFt	.34	

See 0206.2 for Curbs, Gutters, Walks and other Exterior Concrete
See 0302.53 for Topping and Leveling of Floors

	UNIT	COST
TOPPING SLABS		
2" Concrete	SqFt	2.70
3" Concrete	SqFt	3.40
No Mesh or Hoisting		
PADS & PLATFORMS (Including Form Work & Reinforcing)		
4"	SqFt	6.50
6"	SqFt	8.10
PRECAST CONCRETE ITEMS		
Curbs 6" x 10" x 8"	LnFt	11.00
Sills & Stools 6"	LnFt	25.00
Splash Blocks 3" x 16"	Each	60.00
MISCELLANEOUS ADDITIONS TO ABOVE CONCRETE IF NEEDED		
Abrasives - Carborundum - Grits	SqFt	.82
Strips	LnFt	2.70
Bushhammer - Green Concrete	SqFt	1.75
Cured Concrete	SqFt	2.35
Champers - Plastic 3/4"	LnFt	.90
Wood 3/4"	LnFt	.57
Metal 3/4"	LnFt	.93
Colors Dust On	SqFt	.70
Integral (Top 1")	SqFt	1.28
Control Joints - Asphalt 1/2" x 4"	LnFt	.80
1/2" x 6"	LnFt	.91
PolyFoam 1/2" x 4"	LnFt	1.06
Dovetail Slots 22 Ga	LnFt	1.38
24 Ga	LnFt	.90
Hardeners Acrylic and Urethane	SqFt	.24
Epoxy	SqFt	.32
Joint Sealers Epoxy	LnFt	2.54
Rubber Asphalt	LnFt	1.37
Moisture Proofing - Polyethylene 4 mil	SqFt	.16
6 mil	SqFt	.20
Non-Shrink Grouts - Non - Metallic	CuFt	37.00
Aluminum Oxide	CuFt	27.00
Iron Oxide	CuFt	41.00
Reglets Flashing	LnFt	1.48
Sand Blast - Light	SqFt	1.27
Heavy	SqFt	2.30
Shelf Angle Inserts 5/8"	LnFt	4.45
Stair Nosings Steel - Galvanized	LnFt	6.50
Tongue & Groove Joint Forms - Asphalt 3 1/2"	LnFt	1.85
Wood 3 1/2"	LnFt	1.32
Metal 3 1/2"	LnFt	1.75
Treads - Extruded Aluminum	LnFt	8.90
Cast Iron	LnFt	10.10
Water Stops Center Bulb - Rubber - 6"	LnFt	10.30
9"	LnFt	18.00
Polyethylene - 6"	LnFt	4.02
9"	LnFt	4.90

The unit costs in this Division are priced as being done by masonry or General Contractors. See Section 4A for Total Subcontracted Cost. In general, the pricing is based on Modular Units, and materials F.O.B. job site in truckload lots. The basic labor units do include mortar mixing, low scaffolding, tending, and labor fringe benefits. Cleaning, high scaffold, equipment reinforcements, taxes and insurance, etc. must be added to basic labor and material units. The basis for Labor Unit Costs can be found as follows: as to labor rate mean, Page 4-1, and as to number of units placed per day, Table 4B. Unit Costs are based on an average crew of 4 Bricklayers to 3 Tenders (including Mortar Mixer and Operators) for Brick and Stone Work, and 1 Bricklayer to 1 Tender for Block and Tile Work.

		Nominal Dimensions	UNIT	LABOR	MATERIAL (Truckload)
0401.0	**BRICK MASONRY (Bricklayers)**				
.1	FACE BRICK (See Table 4B for Mfg. Size)				
.11	Conventional	8" x 2-2/3" x 4"			
	Running Bond		M pcs	765.00	500.00
	Common Bond (6 Cs. Headers)		M pcs	775.00	500.00
	Stack Bond		M pcs	790.00	500.00
	Dutch & English Bond		M pcs	830.00	500.00
	Flemish		M pcs	830.00	500.00
.12	Economy	8" x 4" x 4"	M pcs	920.00	740.00
.13	Panel (Triple)	8" x 8" x 4"	M pcs	1,500.00	1,650.00
		8" x 16" x 4"	M pcs	2,480.00	3,300.00
.14	Norman	12" x 2-2/3" x 4"	M pcs	900.00	740.00
.15	King	10" x 2-5/8" x 4"	M pcs	830.00	500.00
.16	Norwegian	12" x 3-1/5" x 4"	M pcs	980.00	840.00
.17	Saxon-Utility 3" wall	12"x 4"x 3" (thin)	M pcs	1,190.00	1,000.00
	4" wall	12"x 4"x 4"	M pcs	1,200.00	1,030.00
	6" wall	12"x 4"x 6"	M pcs	1,440.00	1,460.00
	8" wall	16"x 3-5/8"x 3-5/8"	M pcs	1,630.00	1,880.00
.18	Meridian	12" x 2-1/4 " x 4"	M pcs	750.00	1530.00
.19	Adobe Brick - Mexican	8" x 2-3/8"x 4"	M pcs	820.00	620.00
		12"x 2-3/8"x 4"	M pcs	940.00	720.00
.2	COATED BRICK (Ceramic Veneer)		M pcs	850.00	1,120.00
.3	COMMON BRICK - Clay		M pcs	610.00	305.00
	Concrete		M pcs	610.00	275.00
.4	FIRE BRICK - Light Duty (domestic)		M pcs	735.00	1,100.00
	Medium Duty		M pcs	745.00	1,750.00
	Heavy Duty		M pcs	770.00	2,300.00
	Add for Fireplaces		M pcs	600.00	-
	Add: Less than Truckload Lot(14M pcs)				15%
	Add: Full Size Brick		M pcs	4%	8%
	Add: Ceramic Coating - to Common Brick		M pcs	15%	80%
	Deduct: 3" thick Brick (if available)		M pcs	4%	7%
	Add: Ea. Addl. 10' in Fl. Hgt. (or floor)		M pcs	3%	-
	Add: Hauling & Unloading if required		M pcs		See Table 4B
	Add: Breakage & Cutting Allowance		-	-	3%
	Add: Piers & Corbels		M pcs	15%	-
	Add: Sills & Soldiers		M pcs	20%	-
	Add: Floor Brick (over 2")		M pcs	10%	5%
	Add: Weave & Herringbone Pattern		M pcs	20%	7%
	Add: Stack bond		M pcs	8%	-
	Add: Circular or Radius Work		M pcs	20%	-
	Add: Rock Faced		M pcs	10%	-
	Add: Two-Face Joint Striking		M pcs	15%	-
	Add: Arches (including Formwork)		M pcs	75%	20%
	Add: Sawing (Special or Excessive)		M pcs	25%	5%
	Add: Winter Work (below 40 degrees)				
	Production Loss		M pcs	10%	-
	Encls - Wall Area - 1 Side & 3 Uses		Sq Ft	.20	.20
	2 Sides & 3 Uses		Sq Ft	.30	.30
	Heaters & Fuel - Wall Area		Sq Ft	.10	.20
	Heat Mortar		Cu Yd	11.00	5.50
	Deduct: Residential Work		-	10%	-

Crew - 4 Bricklayers to 3 Tenders (Incl. Mortar Mixers & Operators).
See Section 0905.7 for Thin Veneer Brick and Stone (under 2").
See Section 0206.2 for Paver Brick for Sitework.

0402.0 CONCRETE BLOCK (Bricklayers)

		UNIT	LABOR	MATERIAL StdWt	MATERIAL LtWt
.1	CONVENTIONAL				
.11	Foundation or Not Struck - Deduct .35 from Labor Units below.				
.12	Backup or Struck One Face - Deduct .20 from Labor Units below.				
.13	Partition or Struck Two Faces Used for Labor Units below:				
	12" x 8" x 16" Plain	Ea	2.40	1.57	2.10
	Double or Single Corner	Ea	2.43	1.67	2.15
	Bull Nose (Sgl or Dbl)	Ea	2.52	1.85	2.35
	Bond Beam	Ea	2.47	1.75	2.25
	Header	Ea	2.47	1.75	2.25
	12" x 8" x 8" Half Length Block	Ea	2.30	1.58	2.20
	12" x 4" x 16" Half High Block	Ea	2.30	1.61	2.12
	12" x 8" x 8" Lintel	Ea	2.52	1.74	2.24
	12" x 16" x 8" Lintel	Ea	2.68	1.80	2.30
	12" x 8" x 16" Control Joint	Ea	2.52	1.85	2.35
	Half Control Joint	Ea	2.31	1.85	2.45
	Fire Rated (4 Hr.)	Ea	2.41	1.85	2.45
	L Corner	Ea	2.52	1.68	2.17
	Solid or Cap	Ea	2.49	1.90	2.40
	8" x 8" x 16" Plain	Ea	2.25	1.16	1.56
	Double or Single Corner	Ea	2.30	1.26	1.76
	Bullnose (Sgl or Dbl)	Ea	2.32	1.45	1.95
	8" x 8" x 8" Half Bullnose	Ea	2.27	1.45	1.95
	8" x 8" x 16" Bond Beam	Ea	2.32	1.45	1.85
	Header	Ea	2.32	1.45	1.85
	8" x 8" x 8" Half Length Block	Ea	2.22	1.17	1.57
	8" x 8" x 16" Half High Block	Ea	2.22	1.17	1.67
	8" x 8" x 8" Lintel	Ea	2.30	1.37	1.77
	8" x 16" x 8" Lintel	Ea	2.52	1.37	1.77
	8" x 8" x 16" Control Joint	Ea	2.31	1.47	1.87
	8" x 8" x 8" Half Control Joint	Ea	2.20	1.47	1.87
	8" x 8" x 16" Fire Rated (2 Hr.)	Ea	2.25	1.32	1.72
	(4 Hr.)	Ea	2.25	1.47	1.87
	Solid or Cap	Ea	2.30	1.37	1.77
	6" x8" x 16" Plain	Ea	2.16	.97	1.37
	Bull Nose (Sgl or Dbl)	Ea	2.18	1.30	1.70
	6" x8" x 8" Half Bullnose	Ea	2.14	1.30	1.70
	6" x8" x 16" Bond Beam or Lintel	Ea	2.20	1.18	1.58
	6" x8" x 8" Half Length Block	Ea	2.14	1.01	1.41
	6" x4" x 16" Half High Block	Ea	2.14	1.05	1.45
	6" x8" x 16" Fire Rated (2 Hr.)	Ea	2.16	1.14	1.54
	L Corner	Ea	2.33	1.24	1.64
	Solid or Cap	Ea	2.18	1.34	1.74
	4" x 8" x 16" Plain	Ea	2.08	.78	1.18
	Bull Nose	Ea	2.11	1.07	1.37
	Bond Beam or Lintel	Ea	2.13	.96	1.27
	4" x 8" x 8" Half Length Block	Ea	2.03	.80	1.20
	4" x 8"x 16" Half High Block	Ea	2.03	.85	1.15
	L Corner	Ea	2.29	1.00	1.30
	Solid or Cap	Ea	2.15	1.10	1.40

0402.0 CONCRETE BLOCK (Bricklayers)

		UNIT	LABOR	MATERIAL StdWt	MATERIAL LtWt
.13	Partition or Struck Two Faces , Cont'd...				
	16" x 8" x 16" Plain	Ea	2.80	1.90	2.60
	16" x 8" x 8" Half Length	Ea	2.65	1.88	2.58
	14" x 8" x 16" Plain	Ea	2.50	1.87	2.55
	14" x 8" x 8" Half Length	Ea	2.35	1.84	2.53
	14" x 8" x 16" Bond Beam	Ea	2.40	2.10	2.72
	10" x 8" x 16" Plain	Ea	2.30	1.45	1.98
	10" x 8" x 8" Half Length	Ea	2.20	1.42	1.96
	10" x 8" x 16" Bond Beam	Ea	2.33	1.63	2.15
	3" x 8" x 16" Plain	Ea	2.05	.78	1.05
	Solid	Ea	2.15	.82	1.35
	Add or Deduct to all Blockwork Above for:				
	Add: Full Size Block	Ea	.09	.08	-
	Add: Breakage and Cutting	Ea	.10	3%	3%
	Add: Jamb and Sash Block	Ea	.20	.12	.12
	Add: Stack Bond Work	Ea	.20	-	-
	Add: Radius or Circular Work	Ea	1.00	-	-
	Add: Pilaster, Pier or Pedestal Work	Ea	.45	-	-
	Add: Knock-out Block	Ea	.45	.12	.12
	Deduct: Light Weight Block-Installation	Ea	.10	-	-
	Add: Winter Production Loss (below 40°)	Ea	10%	-	-
	Add: Enclosures - Unit	Ea	.20	.09	.09
	Add: Enclosures - Wall Area 1 Side	SqFt	.22	.18	.18
	Add: Enclosures - Wall Area 2 Sides	SqFt	.32	.28	.28
	Add: Heating & Fuel - Wall Area	SqFt	.10	.20	.22
	Deduct: Residential Work	Ea	10%	-	-
	See 0410 for Core Filling for Reinforced Masonry				
	See 0411 for Filling for Insulated Masonry				
.2	ORNAMENTED OR SPECIAL EFFECT				
.21	Shadow wall and Hi-Lite Block				
	Add to Block Prices 0402.1	Ea	.16	.14	.15
.22	Scored Block				
	Add to Block Prices 0402.1 - 1 Score	Ea	.22	.11	.11
	- 2 Score	Ea	.30	.18	.18
.23	Break Off Block				
	Add to Block Prices 0402.1 - Broken	Ea	.20	.65	.65
	- Unbroken	Ea	.28	.60	.60
.24	Split Face Block				
	Add to Block Prices 0402.1	Ea	.20	.28	.30
.25	Rock Faced Block				
	Add to Block Prices 0402.1	Ea	.20	.18	.18
.26	Adobe (or Slump) Block				
	Add to Block Prices 0402.1				
	8" x 8" x 16"	Ea	.23	.32	-
	4" x 8" x 16"	Ea	.22	.22	-
.3	SCREEN WALL - 4" x 12" x 12"	Ea	2.20	1.80	-
.4	INTERLOCKING				
.41	Epoxy Laid				
	Add to Block Prices 0402.1	Ea	-	.15	.15
	Deduct from Labor Costs 0402.1	Ea	.42	-	-
.42	Panelized (No Head Joint Mortar)				
	Add to Block Prices 0402.1	Ea	-	.15	.15
	Deduct from Labor Costs 0402.1	Ea	.50	-	-

0402.0 CONCRETE BLOCK, Cont'd...

		UNIT	LABOR	MATERIAL
.5	SOUND BLOCK			
	Add to Block Prices 0402.1	Ea	.30	.85
	Add for Fillers	Ea	.18	.30
.6	BURNISHED BLOCK - One Face Finish (Truckload Price)			
	Plain 12" x 8" x 16"	Ea	2.70	3.50
	8" x 8" x 16"	Ea	2.55	3.20
	6" x 8" x 16"	Ea	2.45	2.60
	4" x 8" x 16"	Ea	2.35	2.30
	2" x 8" x 16"	Ea	2.25	2.20
	Shapes 12"	Ea	3.15	4.50
	8"	Ea	3.00	3.50
	6"	Ea	2.95	2.90
	4"	Ea	2.85	2.60
	2"	Ea	2.75	2.50
	Add for Bond Beam	Ea	.40	.90
	Add to above for Two Face Finish	Ea	.50	1.75
	Add to above for Scored Face	Ea	.35	.15
	Add for Color	Ea	-	.40
	Add for Bullnose	Ea	.40	.80
.7	PREFACED UNITS - Ceramic Glazed (Truckload Price)			
	12" x 8" x 16" Stretcher	Ea	3.40	6.20
	Glazed 2 Face	Ea	3.95	9.40
	8" x 8" x 16" Stretcher	Ea	3.25	5.70
	Glazed 2 Face	Ea	3.60	9.10
	6" x 8" x 16" Stretcher	Ea	3.30	5.40
	Glazed 2 Face	Ea	3.55	9.00
	4" x 8" x 16" Stretcher	Ea	3.40	5.10
	Glazed 2 Face	Ea	3.65	9.10
	2" x 8" x 16" Stretcher	Ea	3.10	5.10
	4" x 16" x 16" Stretcher	Ea	5.70	17.50
	Add for Scored Block	Ea	.35	.40
	Add for Base Caps, Jambs, Headers	Ea	.63	1.80
	Add for Raked Joints and White Cement	Ea	.75	.15
	Add for Less than Truckload Lot	Ea	-	10%

Crew Ratio - 1 Block Layer to 1 Tender
(Tender includes Mixers & Operators)

Deduct 10% from Block Labor for Residential Construction

0403.0 CLAY BACKING AND PARTITION TILE

(Carload Price)

	UNIT	LABOR	MATERIAL
3" x 12" x 12"	Sq Ft	1.80	1.84
4" x 12" x 12"	Sq Ft	1.90	1.89
6" x 12" x 12"	Sq Ft	2.05	2.35
8" x 12" x 12"	Sq Ft	2.15	2.80

		UNIT	LABOR	MATERIAL
0404.0	**CLAY FACING TILE**			
.1	6T SERIES (SELECT QUALITY - CLEAR & FUNCTIONAL)			
	2" x 5-1/3" x 12" Soap Stretcher (Open Back)	Ea	2.43	3.20
	Soap Stretcher (Solid Back)	Ea	2.50	3.32
	4" x 5-1/3" x 12" 1 Face Stretcher	Ea	2.60	4.45
	Select 2 Face Stretcher	Ea	2.75	5.65
	6" x 5-1/3" x 12" Select 1 Face Bond Stretcher	Ea	2.60	4.87
	8" x 5-1/3" x 12" Select 1 Face Bond Stretcher	Ea	2.75	6.03
	Shapes- Group 1 Square and B/N Corners	Ea	2.82	4.93
	2 Sills and 2" Base	Ea	2.90	5.25
	3 Sills and 4" Base	Ea	2.93	6.47
	4 Starters and Octagons	Ea	3.30	12.02
	5 Radials and End Closures	Ea	3.20	20.00
	Add for Less than Carload Lots	Ea	-	10%
	Add for Designer Colors	Ea	-	20%
	Add for Base Only	Ea	-	40%
.2	8W SERIES (SELECT QUALITY - CLEAR & FUNCTIONAL)			
	2" x 8" x 16" Soap Stretcher (Open Back)	Ea	3.28	5.35
	Soap Stretcher (Solid Back)	Ea	3.36	5.76
	4" x 8" x 16" 1 Face Stretcher	Ea	3.35	5.98
	Select 2 Face Stretcher	Ea	3.55	9.15
	6" x 8" x 16" Select 2 Face 6" Bond Stretcher	Ea	3.50	7.85
	8" x 8" x 16" Select 1 Face 8" Bond Stretcher	Ea	3.65	9.55
	Shapes- Group 1 Square and B/N Corners	Ea	3.60	8.30
	2 Sills and 2" Base	Ea	4.10	10.70
	3 Sills and 4" Base	Ea	4.20	13.30
	4 Starters and Octagons	Ea	4.55	16.30
	5 Radials and End Closures	Ea	4.65	40.00
	Add for Less than Carload Lots	Ea	-	40%
	Add for Designer Colors	Ea	-	20%
	Add for Base Only	Ea	25%	-
0405.0	**GLASS UNITS**			
	4" x 8" x 4"	Ea	3.50	5.25
	6" x 6" x 4"	Ea	3.80	5.30
	8" x 8" x 4"	Ea	4.35	6.20
	12" x 12" x 4"	Ea	5.60	17.00
	Add for Solar Reflective Type	Ea	-	40%
	Accessories: Reinforcing & Expansion Joint	Ea	-	2.00
0406.0	**TERRA-COTTA**			
	Unglazed	SqFt	2.50	5.80
	Glazed	SqFt	2.90	6.90
	Colored Glaze	SqFt	3.00	10.50
0407.0	**MISCELLANEOUS UNITS**			
.1	COPING TILE - 9" Double Slant	LnFt	3.80	8.70
	12" Double Slant	LnFt	4.15	11.50
	Bell	LnFt	4.15	9.50
	Corners & Ends = 4 x Above Material Prices			
.2	FLUE LINING - 8" x 8"	LnFt	3.60	5.00
	8" x 12"	LnFt	3.85	9.00
	12" x 12"	LnFt	4.25	10.50
	16" x 16"	LnFt	7.50	16.00
	18" x 18"	LnFt	8.60	18.00
	20" x 20"	LnFt	14.30	34.00
	24" x 24"	LnFt	19.00	41.00

0408.0 NATURAL STONE (M) (Bricklayers)

(See 0906.0 for Stone under 2")

		SQUARE FOOT Labor	SQUARE FOOT Matl.	CUBIC FOOT Labor	CUBIC FOOT Matl.
.1	CUT STONE				
.11	Limestone				
	Indiana (Standard) and Alabama				
	Face Stone 3"	8.50	17.30	34.00	69.20
	4"	9.50	18.80	28.50	56.40
	Sills and Light Trim	-	-	36.80	67.00
	Copings and Heavy Sections	-	-	33.50	65.00
	Add for Select	1.50	-	6.00	
	Minnesota, Texas, Wisconsin, etc.				
	Face Stone 3"	8.50	22.50	34.00	90.00
	4"	9.60	24.50	28.80	73.50
	Sills and Light Trim	-	-	28.88	89.20
	Copings and Heavy Sections	-	-	27.55	87.00
.12	Marble				
	Face Stone 2"	8.70	28.00	52.20	168.00
	3" and 4"	9.50	30.00	36.00	105.00
	Sills and Light Trim	-	-	37.00	110.00
	Copings and Heavy Sections	-	-	33.00	93.00
.13	Granite				
	Face Stone 2"	8.90	25.00	53.40	150.00
	3"	10.75	27.00	43.00	108.00
	4"	12.50	31.00	37.50	93.00
	Sills and Light Trim	-	-	44.50	102.00
	Copings and Heavy Sections	-	-	41.25	98.00
.14	Slate 1 1/2"	9.50	27.00	57.00	162.00
.2	ASHLAR STONE (40 SqFt/Ton - 4" sawed bed)				
.21	Limestone				
	Indiana - Random	8.85	9.30		
	Coursed (2"-5"-8")	9.20	10.50		
	Minn., Ala., Texas, Wisc., etc.)				
	Split Face-1 Size 7 1/2"	9.45	10.30		
	Coursed (2"-5"-8")	11.00	11.30		
	Sawed/Planed Face-Random	9.10	11.40		
	Coursed (2"-5"-8")	8.70	12.60		
.22	Marble- Sawed Random	9.80	18.35		
	Coursed	10.10	18.90		
.23	Granite- Bushhammered-Random	10.35	21.00		
	Coursed	10.75	22.00		
.3	ROUGH STONE (30 s.f. per Ton)				
.31	Flagstone and Rubble 2"	9.80	8.40		
.32	Field Stone or Boulders	9.10	7.30		
.33	Lt.Wt. Boulders (Igneous) (260 s.f. Ton)				
	2" to 5" Veneer (Sawed Back)	7.85	7.35		
	3" to 10" Boulders	9.30	7.20		
.34	Lt.Wt. Limestone (90 s.f. Ton)				
	2" to 4"	8.85	9.00		

0409.0 PRECAST VENEERS AND SIMULATED MASONRY

		Labor	Matl.
.1	ARCHITECTURAL PRECAST STONE		
	Limestone and Gravel	8.10	18.30
	Marble	8.90	25.60
	Granite and Quartz	8.95	23.70
.2	PRECAST CONCRETE	7.80	14.10
.3	MOSAIC GRANITE PANELS	11.20	29.40

		UNIT	LABOR	MATERIAL
0410.0	**MORTARS**			
	(See Table 4B for Quantities per Unit)			
	Material Cost used for Units Below (Truckload Prices)			
	Portland Cement - Gray	Bag- 94#	10.60	9.50
	White	Bag- 94#	22.50	18.60
	Masonry Cement	Bag- 70#	10.15	7.30
	High Early Cement	Bag- 94#	-	10.50
	Lime, Hydrated	Bag- 50#	6.35	5.25
	Silica Sand	Bag-100#	-	7.00
	Fire Clay	Bag-100#	-	20.50
	Sand	CuYd	-	21.00
	Add for Less than Truckload	-	-	20%
	Labor Incl. in Unit Costs: with Mortar Mixer	CuYd	26.00	-
	Hand Mixed	CuYd	60.00	-
.1	PORTLAND CEMENT AND LIME MIX			
	(4.5 Bags Portland and 4.5 Bags Lime)			
	1:1:6 - Gray - Truckload	CuYd	26.30	90.00
	Less than Truckload	CuYd	26.30	108.00
	White	CuYd	26.30	128.00
	Add for Using Silica Sand	CuYd	22.20	39.00
.2	MASONRY CEMENT MIX (9 Bags Cement)			
	1:3 - Truckload	CuYd	26.50	86.70
	Less than Truckload	CuYd	26.50	96.60
.3	EPOXY CEMENT	CuYd	34.60	178.00
.4	FIRE RESISTANT CEMENT (Clay Mix 300# Mpc)	CuYd	34.60	98.50
.5	ADMIXTURES (Add to Mortar Prices above)			
	Waterproofing (1# sack cmt & $ .40/lb)	CuYd	18.00	9.00
	Accelerators (1 qt sack cmt & $1.00/lb)	CuYd	18.00	12.60
.6	COLORING (50# CuYd or 90# Mpc & $1.00/lb)	CuYd	20.70	48.35
	Standard Brick	M Pcs	25.00	99.20
	Add for Heating Mortar Materials	CuYd	11.50	7.25
	10% allowed in above Prices for Shrinkage			

0411.0 CORE FILLING FOR REINFORCED CONCRETE MASONRY

Based on 3500# Conc.@ $76 CuYd; Mortar @ $90 CuYd; and $26 CuYd to Mix
Deduct 50% from Labor Units if Truck Chuted or Pumped

	Void			Ready Mix		Job Mix	
	CuFt	%	Unit	L	M	L	M
12" x 8" x 16" Plain - 2 Cell	.43	48	Pc	.22	1.21	.38	1.77
	.48	53	SqFt	.25	1.34	.40	1.98
8" x 8" x 16" Plain - 2 Cell	.27	40	Pc	.14	.76	.23	1.10
	.30	44	SqFt	.15	.84	.25	1.24
6" x 8 " x16" Plain - 2 Cell	.19	40	Pc	.10	.54	.34	.78
	.21	44	SqFt	.11	.59	.36	.86
12" x 8" x 16" Bond Beam	.45	50	Pc	.24	1.26	.64	1.85
	.49	55	SqFt	.26	1.37	.75	2.02
8" x 8" x 16" Bond Beam	.21	32	Pc	.11	.59	.18	.86
	.23	35	SqFt	.12	.65	.19	.95
6" x 8" x 16" Bond Beam	.11	25	Pc	.06	.31	.10	.45
	.13	27	SqFt	.07	.37	.11	.53
12" x 16" x 8" Lintel	.52	58	pc	.27	1.46	.44	2.14
	.57	63	SqFt	.30	1.60	.48	2.35
8" x 16" x 8" Lintel	.27	40	Pc	.21	.76	.23	1.11
	.30	44	SqFt	.16	.84	.25	1.23
16" x 8" x 18" Pilaster Block	.96	72	Pc	.50	2.68	.80	3.96
	1.08	81	LnFt	.56	3.00	.90	4.46
Add for Vertical Rein.			Pc	.25	.25	.25	.25
(1 #4 Rods 16" O.C.)			SqFt	.26	.26	.26	.26

0412.0 CORE & CAVITY FILL FOR INSULATED MASONRY

		R Value-No Insulation	R Value-Insulated	UNIT	LABOR	MATERIAL
.1	LOOSE					
.11	Core Fill					
	12" Concrete Block	1.28	3.90	Pc	.34	.42
				SqFt	.37	.45
	Lt. Wt. Concrete Block	2.27	8.30	Pc	.34	.42
				SqFt	.37	.45
	10" Concrete Block	1.20	2.40	Pc	.30	.36
				SqFt	.32	.39
	Lt. Wt. Concrete Block	2.22	5.70	Pc	.29	.37
				SqFt	.32	.41
	8" Concrete Block	1.11	1.93	Pc	.23	.24
				SqFt	.24	.25
	Lt. Wt. Concrete Block	2.18	5.03	Pc	.23	.23
				SqFt	.24	.24
	6" Concrete Block	.90	1.79	Pc	.21	.17
				SqFt	.22	.18
	Lt. Wt. Concrete Block	1.65	2.99	Pc	.21	.17
				SqFt	.22	.18
			R Value			
	8" Brick		.88	SqFt	.22	.13
	6" Brick		.66	SqFt	.18	.09
	4" Brick		.44	SqFt	.16	.07
	Add to above for Expanded Mica		-	-	10%	200%
	Add to above for Perlite		-	-	-	100%
	10% added for spillage					
.12	Cavity Fill - Per Inch Wall					
	Expanded Styrene		3.8	SqFt	.23	.20
	Fiber Glass		2.2	SqFt	.23	.23
	Rock Wool		2.9	SqFt	.23	.23
	Cellulose		3.7	SqFt	.23	.21
	Add per added inch		-	-	40%	100%
	10% added for spillage					
.2	RIGID					
	Fiber Glass					
	1" - 3# Density		4.35	SqFt	.31	.33
	1.5"		6.52	SqFt	.34	.50
	2"		8.70	SqFt	.38	.65
	Expanded Styrene - Molded (White)					
	1" - 1# Density		4.33	SqFt	.31	.14
	1.5"		6.52	SqFt	.34	.21
	2"		8.70	SqFt	.38	.28
	Add for Embedded Nailer Type		-	-	-	.11
	Expanded Styrene - Extruded (Blue)					
	1" - 2# Density		5.40	SqFt	.31	.38
	1.5"		8.10	SqFt	.34	.17
	2"		10.80	SqFt	.38	.66
	Expanded Urethane					
	1"		7.14	SqFt	.31	.40
	2"		14.28	SqFt	.38	.58
	Perlite					
	1"		2.78	SqFt	.31	.37
	2"		5.56	SqFt	.38	.74
	Add for Glued Applications		-	SqFt	.05	.02

0413.0 ACCESSORIES & SPECIALTIES - MATERIAL

		UNIT	LABOR	MATERIAL
.1	ANCHORS			
	Dovetail-Brick 12 ga Galvanized x 3 - 1/2"	Ea	.19	.44
	Brick 16 ga Galvanized x 3 - 1/2"	Ea	.18	.29
	Brick 16 ga Galvanized x 5"	Ea	.19	.37
	Stone - Stainless	Ea	.19	1.10
	Rigid Wall - 1 - 1/4"x 3/6"x 8" - Galv	Ea	.18	.97
	Stic - Kips and Clamp	Ea	.23	.20
	Strap Z - 1/8"x 1"x 6" or 8" - Galv	Ea	.38	1.15
	1/8" x 1"x 6" - SS	Ea	.39	2.50
	3/16" x 1 - 1/4" x 8" - Galv	Ea	.40	1.65
	Split Tail - 3/16" x 1 - 1/4" x 6" - Galv	Ea	.38	1.60
	1/8" x 1 - 1/2"x 6" - SS	Ea	.39	3.35
	Wall Clips - 2"x 3"	Ea	.38	.20
	Wall Plugs - Wood Filled	Ea	.38	.22
.2	TIES - WIRE			
	Galv Z - 3/16" Wire - 6" & 8"	Ea	.18	.17
	Zinc Z - 3/16" Wire - 6" & 8"	Ea	.18	.40
	Copperweld Z - 3/16" Wire - 6"	Ea	.18	.45
	3/16" Wire - 8"	Ea	.18	.52
	Adjustable Rect. - 3 /16" Wire 3 - 1/4"	Ea	.18	.30
	3/16" Wire 4 - 3/4"	Ea	.18	.35
	Corrugated-Strap - 26 ga - 7/8" x 7"	Ea	.18	.04
	22 ga - 7/8" x 7"	Ea	.19	.05
	16 ga - 7/8" x 7"	Ea	.20	.08
	Hardware Cloth - 2" x 5" x 1/4" - 23 ga	Ea	.17	.14
	23 ga x 1/4" x 30"	SqFt	.25	.55
.3	REINFORCEMENT			
	#9 Welded Wire - Ladder Type - 4"	LnFt	.17	.15
	6"	LnFt	.17	.16
	8"	LnFt	.18	.18
	10"	LnFt	.18	.19
	12"	LnFt	.20	.20
	16"	LnFt	.22	.28
	Add for 3/16" Wire H.D.	LnFt	-	30%
	Add for Truss Type	LnFt	-	20%
	Add for Rect. Ties - Composite & Cavity	LnFt	.03	.06
	Add for Rect. Adj. Ties - Comp & Cavity	LnFt	.03	.06
	Reinforcement Rods - Horizontal Placement			
	3/8" or #3 .376 lb	LnFt	.18	.21
	1/2" or #4 .668 lb	LnFt	.23	.27
	5/8" or #5 1.043 lb	LnFt	.32	.34
	Add for Vertical Placement	LnFt	25%	-
.4	FIREPLACE AND CHIMNEY ACCESSORIES			
	Ash Dumps and Thimbles - 6" x 8"	Ea	29.00	12.50
	Clean Cut Doors, Cast Iron - 8" x 8"	Ea	27.00	21.00
	12" x 12"	Ea	32.20	44.00
	Dome Dampers, Cast Iron - 24" x 16"	Ea	24.10	50.00
	30" x 16"	Ea	28.00	65.00
	36" x 20"	Ea	42.00	93.00
	42" x 20"	Ea	55.00	98.00
	48" x 24"	Ea	76.50	135.00

		UNIT	LABOR	MATERIAL 1/2"	MATERIAL 1"
0413.0	**ACCESSORIES & SPECIALTIES, Cont'd...**				
.5	CONTROL AND EXPANSION JOINT MATERIAL				
	Asphalt - Fibre	SqFt	.56	1.35	2.30
	Polyethylene Foam	SqFt	. 57	1.40	2.40
	Sponge Rubber	SqFt	. 57	6.50	8.00
	Paper Fibre	SqFt	. 55	.90	1.60
	Also in Thk. of 1/8",1/4",3/8" & 3/4"				
.6	FLASHING, THRU WALL			2 oz	3 oz
	Copper Between Polyethylene	SqFt	.80	1.25	1.55
	Rein Paper - 2 oz/1.05	SqFt	.75	1.40	2.12
	Asphalted Fabric - 3 oz	SqFt	.75	1.90	2.60
	Poly Vinyl Chloride - 20 mil	SqFt	.60	-	.24
	30 mil	SqFt	.65	-	.40
	Add for Adhesives			.05	.05
0414.0	**CLEANING & POINTING** (See 0414 for Restoration)				
.1	BRICK & STONE- Acid & Other Chemicals	SqFt	.43	-	.03
	Soap & Water	SqFt	.38	-	-
.2	BLOCK - 1 Face	SqFt	.22	-	-
.3	FACING TILE AND BLOCK				
	Point - White Cement and Sand	SqFt	.75	-	.08
	Silica Sand	SqFt	.92	-	.13
	Add for Scaffold (See 0417)				

		UNIT	Cost
0415.0	**MASONRY RESTORATION (L&M) (Bricklayers)**		
.1	RAKING, FILLING AND TUCKPOINTING		
	Limited - Standard Brick & 3/8" Joint	SqFt	3.05
	Heavy	SqFt	3.80
	All	SqFt	7.40
	Deduct for Stone	SqFt	20%
	Deduct for 4" x 12" Brick	SqFt	30%
	Add for Larger & Smaller Joints	SqFt	25%
.2	SAND BLAST		
.22	Brick -Wet	SqFt	1.70
	Dry	SqFt	1.58
.23	Block -Wet	SqFt	1.58
	Dry	SqFt	1.45
.3	STEAM CLEAN AND CHEMICAL CLEANING	SqFt	1.42
.4	HIGH PRESSURE WATER CLEANING (Hydro)	SqFt	1.00
.5	BRICK REPLACEMENT -Standard	Each	20.00
	4" x 12"	Each	25.00
	Add for Scaffolding	SqFt	.70
	Add for Protection	SqFt	.30
.6	WATER PROOFING - CLEAR	SqFt	.85
	See 0702.3 for Waterproofing Masonry		
	See 0711.0 for Caulking		
0416.0	**INSTALLATION OF STEEL & MISC. EMBEDDED ITEMS**		
	See Division 0502		

		UNIT	LABOR	MATERIAL
0417.0	**EQUIPMENT AND SCAFFOLD**			
.1	EQUIPMENT AVERAGE OF LABOR			
	See 1-7A for Rental Rates			
.2	SCAFFOLD - INCLUDING PLANK			
	Tubular Frame- to 40' - Exterior	SqFt	.40	.22
	to 16' - Interior	SqFt	.37	.23
	Swing Stage (Incl. Outriggers & Anchorage)	SqFt	.36	.25

The costs below are average but can vary greatly with Quantity, Weather, Area Practices and many other factors. They are priced as total contractor's price with an overhead and fee of 10% included. Units include mortar, ties, cleaning, and reinforcing. Not included are scaffold, hoisting, lintels, accessories, heat, and enclosures. Also added are 27% Insurance on Labor; 10% for General Conditions, Equipment and Tools; and 5% on Material for Sales Taxes. Crew Size is based on 4 Bricklayers to 3 Tenders for Brick and Stone Work, and 4 Bricklayers to 4 Tenders for Block and Tile. See Table 3 for Production Standards. Deduct 5% for Residential Work.

Example:			UNIT	LABOR	MATERIAL	TOTAL
.1	Face Brick 8" x 2 2/3" x 4"					
.11	Running Bond		Each	.76	.50	1.26
	Mortar and Clean Brick		Each	.10	.10	.20
				.86	.60	1.46
	Add: 27% Labor Burden, 5% Sales Tax			.24	.03	.27
				1.10	.63	1.73
	Add: 10% General Conditions & Equipment					.17
						1.90
	Add: 10% Overhead and Fee					.19
	Total Brick Cost			Each Unit		2.09
				or SqFt		14.10

0401.0 BRICK MASONRY

		COST SqFt	COST Unit
.1	FACE BRICK		
.11	Conventional (Modular Size)		
	Running Bond - 8" x 2 2/3" x 4"	14.10	2.09
	Common Bond - 6 Course Header	16.50	2.10
	Stack Bond	14.30	2.12
	Dutch & English Bond - Every Other Course Header	22.20	2.25
	Every Course Header	25.25	2.25
	Flemish Bond - Every Other Course Header	17.00	2.25
	Every Course Header	20.25	2.25
	Add: If Scaffold Needed	.75	.12
	Add: For each 10' of Floor Hgt. or Floor (3%)	.40	.05
	Add: Piers and Corbels (15% to Labor)	1.21	.18
	Add: Sills and Soldiers (20% to Labor)	1.43	.22
	Add: Floor Brick (10% to Labor)	.84	.12
	Add: Weave & Herringbone Patterns (20% to Labor)	1.48	.22
	Add: Stack Bond (8% to Labor)	.67	.10
	Add: Circular or Radius Work 20% to Labor)	1.48	.22
	Add: Rock Faced & Slurried Face (10% to Labor)	.84	.12
	Add: Arches (75% to Labor)	4.32	.64
	Add: For Winter Work (below 40°)		
	Production Loss (10% to Labor)	.75	.12
	Enclosures - Wall Area Conventional	.75	.12
	Heat and Fuel - Wall Area Conventional	.30	.04
.12	Econo - 8" x 4" x 3"	11.55	2.57
.13	Panel - 8" x 8" x 4"	10.25	4.56
.14	Norman - 12" x 2 2/3" x 4"	11.45	2.55
.15	King Size - 10" x 2 5/8" x 4"	10.20	2.18
.16	Norwegian - 12" x 3 1/5" x 4"	10.40	2.78
.17	Saxon-Utility - 12" x 4" x 3"	9.96	3.32
	12" x 4" x 4"	10.00	3.35
	12" x 4" x 6"	12.72	4.24
	12" x 4" x 8"	15.40	5.13
.18	Adobe - 12" x 3" x 4"	17.07	2.53
.2	COATED BRICK (Ceramic)	19.97	2.93
.3	COMMON BRICK (Clay, Concrete and Sand Lime)	10.80	1.60
.4	FIRE BRICK -Light Duty	17.80	2.70
	Heavy Duty	21.00	3.15
	Deduct for Residential Work - All Above	-	5%

		UNIT	LABOR	MATERIAL	TOTAL
Example:	12" x 8" x 16" Concrete Block - Partition	Each	2.40	1.57	3.97
	Mortar (Labor in Block Placing Above)	Each	-	.22	.22
	Reinforcing Wire 16" O.C. Horizontal	Each	.10	.12	.22
	Clean Block - 2 Faces	Each	.35	-	.35
			2.85	1.91	4.76
	Add: 27% Taxes and Insurance on Labor				.77
	Add: 5% Sales Tax on Material				.10
					5.63
	Add: 10% General Cond., Equip. and Tools				.57
					6.20
	Add: 10% Overhead and Profit				.62
	Total Block Cost	Each			6.82
		or SqFt			7.67

Not Included - Scaffold, Hoisting Equipment, Lintels, Embedded Accessories, Insulation, Heat and Enclosures. (See "Adds" Below.)

0402.0 CONCRETE BLOCK

		Std Wt Cost		Lt Wt Cost	
.1	CONVENTIONAL (Struck 2 Sides - Partitions)	SqFt	Unit	SqFt	Unit
	12" x 8"x 16" Plain	7.67	6.82	8.40	7.48
	Bond Beam (with Fill-Reinforcing)	10.21	9.08	10.75	9.55
	12" x 8"x 8" Half Block	7.65	6.80	7.25	6.45
	Double End - Header	7.87	7.00	8.44	7.50
	8" x 8" x 16" Plain	6.75	6.00	7.30	6.50
	Bond Beam (with Fill-Reinforcing)	8.17	7.27	8.62	7.67
	8" x 8" x 8" Half Block	6.63	5.90	6.70	5.95
	Double End - Header	6.97	6.20	7.57	6.73
	6 "x 8" x 16" Plain	6.25	5.55	6.81	6.06
	Bond Beam (with Fill-Reinforcing)	7.10	6.32	7.56	6.72
	Half Block	6.21	5.52	6.72	5.98
	4" x 8" x 16" Plain	6.08	5.40	5.60	5.00
	Half Block	5.35	4.76	5.91	5.26
	16" x 8" x 16" Plain	8.70	7.74	9.52	8.45
	Half Block	8.41	7.48	9.20	8.18
	14" x 8" x 16" Plain	8.30	7.40	9.30	8.27
	Half Block	7.87	7.00	8.23	7.32
	10" x 8" x 16" Plain	7.20	6.40	7.94	7.06
	Bond Beam (with Fill-Reinforcing)	8.26	7.35	9.45	8.40
	Half Block	6.92	6.15	7.48	6.65
	Deduct: Block Struck or Cleaned One Side	.25	.23	.25	.23
	Deduct: Block Not Struck or Cleaned Two Sides	.40	.35	.47	.35
	Deduct: Clean One Side Only	.18	.17	.19	.17
	Deduct: Lt. Wt. Block (Labor Only)	.16	.15	.16	.15
	Add: Jamb and Sash Block	.42	.38	.42	.38
	Add: Bullnose Block	.63	.56	.63	.56
	Add: Pilaster, Pier, Pedestal Work	.69	.62	.69	.62
	Add: Stack Bond Work	.27	.24	.27	.24
	Add: Radius or Circular Work	1.24	1.12	1.24	1.12
	Add: If Scaffold Needed	.71	.63	.71	.63
	Add: Winter Production Cost (10% Labor)	.40	.40	.40	.36
	Add: Winter Enclosing - Wall Area	.40	.40	.40	.36
	Add: Winter Heating - Wall Area	.37	.37	.37	.33
	Add: Core & Beam Filling - See 0411 (Page 4A-18)				
	Add: Insulation - See 0412 (Page 4A-18)				

Deduct for Residential Work - 5%

		Std Wt Cost		Lt Wt Cost	
0402.0	**CONCRETE BLOCK, Cont'd...**	SqFt	Unit	SqFt	Unit
.2	ORNAMENTAL OR SPECIAL EFFECT - Add to Units Above in 0402.1				
	Shadow Wall	.45	.40	.47	.42
	Scored Block	.51	.45	.51	.46
	Breakoff	1.26	1.12	1.15	1.02
	Split Face	.73	.65	.78	.69
	Rock Faced	.71	.63	.76	.68
	Adobe (Slump)	.76	.69	-	-
.3	SCREEN WALL - 4" x 12" x 12"	5.80	5.15	-	-
.4	INTERLOCKING - Deduct from Units 0402.1				
	Epoxy Laid	.49	.44	.45	.44
	Panelized	.72	.64	.67	.64
.5	SOUND BLOCK - Add to Block Prices 0402.1			1.71	1.52
.6	BURNISHED - 12" x 8" x 16"			10.46	9.30
	8" x 8" x 16"			9.73	8.65
	6" x 8" x 16"			8.77	7.80
	4" x 8" x 16"			8.15	7.25
	2" x 8" x 16"			7.70	6.85
	Add for Shapes			2.25	2.00
	Add for 2 Faced Finish			3.30	2.95
	Add for Scored Finish - 1 Face			.78	.70
.7	PREFACED UNITS -12" x 8" x 16" Stretcher			15.30	13.60
	(Ceramic Glazed) 12" x 8" x 16" Glazed 2 Face			20.85	18.55
	8" x 8" x 16" Stretcher			14.30	12.70
	Glazed 2 Face			19.70	17.50
	6" x 8" x 16" Stretcher			12.60	11.20
	Glazed 2 Face			19.00	16.90
	4" x 8" x 16" Stretcher			13.50	12.00
	Glazed 2 Face			18.65	16.60
	2" x 8" x 16" Stretcher			13.15	11.70
	4" x 16" x 16" Stretcher			34.40	30.60
	Add for Base, Caps, Jambs, Headers, Lintels			3.82	3.40
	Add for Scored Block			1.12	1.00

		COST	
0403.0	**CLAY BACKING AND PARTITION TILE**	SqFt	Unit
	3" x 12" x 12"	5.07	5.07
	4" x 12" x 12"	5.40	5.40
	6" x 12" x 12"	6.40	6.40
	8" x 12" x 12"	7.10	7.10
0404.0	**CLAY FACING TILE (GLAZED)**		
.1	6T or 5" x 12" - SERIES		
	2" x 5 1/3" x 12" Soap Stretcher (Solid Back)	18.45	8.20
	4" x 5 1/2" x 12" 1 Face Stretcher	19.60	8.70
	2 Face Stretcher	23.20	10.30
	6" x 5 1/3" x 12" 1 Face Stretcher	24.30	10.80
	8" x 5 1/3" x 12" 1 Face Stretcher	27.20	12.10
.2	8W or 8" x 16" - SERIES		
	2" x 8" x 16" Soap Stretcher (Solid Back)	14.10	12.50
	4" x 8" x 16" 1 Face Stretcher	16.50	12.90
	2 Face Stretcher	19.10	16.90
	6" x 8" x 16" 1 Face Stretcher	17.80	15.75
	8" x 8" x 16" 1 Face Stretcher	18.20	16.10
	Add for Shapes (Average)	50%	50%
	Add for Designer Colors	-	20%
	Add for Less than Truckload Lots	-	10%
	Add for Base Only	-	25%

		COST SqFt	COST Unit
0405.0	**GLASS UNITS**		
	4" x 8" x 4"	46.80	12.20
	6" x 6" x 4"	50.80	12.70
	8" x 8" x 4"	31.50	14.05
	12" x 12" x 4"	29.40	29.40
0406.0	**TERRA-COTTA**		
	Unglazed	10.50	10.50
	Glazed	14.80	14.80
	Colored Glazed	17.30	17.30

		Ln Ft
0407.0	**FLUE LINING**	
	8" x 12"	18.05
	12" x 12"	19.25
	16" x 16"	30.80
	18" x 18"	32.30
	20" x 20"	62.50
	24" x 24"	80.30

		Sq Ft	Cu Ft
0408.0	**NATURAL STONE**		
.1	CUT STONE		
	Limestone- Indiana and Alabama - 3"	35.40	141.60
	4"	38.30	114.90
	Minnesota, Wisc, Texas, etc.- 3"	40.50	162.00
	4"	46.40	139.20
.12	Marble - 2"	49.80	298.80
	3"	51.50	206.00
.13	Granite - 2"	46.75	280.50
	3"	50.40	201.60
.14	Slate - 1 1/2"	46.80	280.80
.2	ASHLAR- 4" Sawed Bed		
.21	Limestone- Indiana - Random	25.50	76.50
	Coursed - 2" - 5" - 8"	26.80	80.40
	Minnesota, Alabama, Wisconsin, etc.		
	Split Face - Random	27.60	82.80
	Coursed	30.50	91.50
	Sawed or Planed Face - Random	29.00	87.00
	Coursed	30.60	91.80
.22	Marble - Sawed - Random	38.30	114.90
	Coursed	39.40	118.20
.23	Granite - Bushhammered - Random	40.50	121.50
	Coursed	43.20	129.60
.24	Quartzite	29.00	87.00
.3	ROUGH STONE		
.31	Rubble and Flagstone	26.80	80.40
.32	Field Stone or Boulders	26.60	79.80
.33	Light Weight Boulders (Igneous)		
	2" to 4" Veneer - Sawed Back	22.30	-
	3" to 10" Boulders	23.40	-
.34	Light Weight Limestone	24.65	-

		Unit	Cost
0409.0	**PRECAST VENEERS AND SIMULATED MASONRY**		
.1	ARCHITECTURAL PRECAST STONE- Limestone	SqFt	34.00
	Marble	SqFt	40.50
.2	PRECAST CONCRETE	SqFt	26.90
.3	MOSAIC GRANITE PANELS	SqFt	38.80
	See 0904 for Veneer Stones (under 2")		
0410.0	**MORTAR**		
	Portland Cement and Lime Mortar - Labor in Unit Costs	CuYd	102.00
	Masonry Cement Mortar - Labor in Unit Costs	CuYd	90.00
	Mortar for Standard Brick - Material Only	SqFt	.40

		COST	
		SqFt	Unit
0411.0	**CORE FILLING FOR REINFORCED CONCRETE BLOCK**		
	Add to Block Prices in 0402.0 (Job Mixed)		
	12" x 8" x 16" Plain (includes #4 Rod 16" O.C. Vertical)	3.37	2.76
	Bond Beam (includes 2 #4 Rods)	3.66	3.26
	8" x 8" x 6" Plain (includes #4 Rod 16" O.C. Vertical)	1.89	1.68
	Bond Beam (includes 2 #4 Rods)	1.50	1.34
	6" x 8" x 16" Plain	1.66	1.48
	Bond Beam (includes 1 #4 Rod)	.85	.75
	10" x 8" x 16" Bond Beam (includes 2 #4 Rods)	3.28	2.92
	14" x 8" x 16" Bond Beam (includes 2 #4 Rods)	3.80	3.40
0412.0	**CORE AND CAVITY FILL FOR INSULATED MASONRY**		
	See Page 4-11 for R Values		
.1	LOOSE		
	Core Fill - Expanded Styrene		
	12" x 8" x 16" Concrete Block	1.20	1.07
	10" x 8" x 16" Concrete Block	1.05	.94
	8" x 8" x 16" Concrete Block	.76	.68
	6" x 8" x 16" Concrete Block	.64	.57
	8"Brick - Jumbo - Thru the Wall	.56	.50
	6"Brick - Jumbo - Thru the Wall	.45	.40
	4"Brick - Jumbo	.38	.34
	Cavity Fill - per Inch		
	Expanded Styrene	.50	-
	Mica	.85	-
	Fiber Glass	.52	-
	Rock Wool	.52	-
	Cellulose	.50	-
.2	RIGID - Fiber Glass - 3# Density, 1"	.90	-
	2"	1.40	-
	Expanded Styrene - Molded - 1# Density, 1"	.65	-
	2"	.95	-
	Extruded - 2# Density, 1"	.92	-
	2"	1.40	-
	Expanded Urethane - 1"	1.02	-
	2"	1.32	-
	Perlite - 1"	.98	-
	2"	1.50	-
	Add for Embedded Water Type	.18	-
	Add for Glued Applications	.06	-
0413.0	**ACCESSORIES AND SPECIALTIES**		
	See Page 4-12		
0414.0	**CLEANING AND POINTING**		
	Brick and Stone - Acid or Chemicals	.72	-
	Soap and Water	.65	-
	Block & Facing Tile - 1 Face (Incl. Hollow Metal Frames)	.35	-
	Point with White Cement	1.40	-
0415.0	**MASONRY RESTORATION** - See 0415.0		
0416.0	**INSTALLATION OF STEEL** - See 0502.0		
0417.0	**SCAFFOLD**		
.1	EQUIPMENT, TOOLS AND BLADES		
	Percentage of Labor as an average	-	7%
	See 1 - 7A for Rental Rates & New Costs		
.2	SCAFFOLD- Tubular Frame - to 40 feet - Exterior	.90	-
	Tubular Frame - to 16 feet - Interior	.85	-
	Swing Stage - 40 feet and up	.95	-

TABLE B

		Modular Manufactured Size - Inches (l x h x d)	Mortar CuFt M	Unit Per SqFt	Units Placed ManDay	Haul & Unload M PCs	Unit Wt.
0401.0	**BRICK MASONRY**						
.1	Face Brick						
.11	Conventional						
	Running Bond	7-5/8 x 2-1/4 x 3-5/8	13	6.75	600	16.00	4.5
	Common Bond						
	Header Every 6th Course			ƒ1/6	570	-	-
	Header Every 7th Course			ƒ1/7	570	-	-
	Dutch and English Bond						
	Header Every Other Course			ƒ1/2	510	-	-
	Header Every Course			ƒ2/3	510	-	-
	Flemish Bond						
	Header Every Other Course			ƒ1/8	500	-	-
	Header Every Course			ƒ1/3	500	-	-
.12	Econo	7-5/8 x 3-5/8 x 3-5/8		4.50	430	20.00	6.0
.13	Panel	7-5/8 x 7-5/8 x 3-5/8		2.25	280	40.00	13.5
		7-5/8 x 15-5/8 x 3-5/8		1.25	180	80.00	27.0
.14	Norman	11-5/8 x 2-1/4 x 3-5/8	16	4.50	450	20.00	6.0
.15	King	9-5/8 x 3-1/2 x 3-5/8	16	4.70	440	22.00	4.5
.16	Norwegian	11-5/8 x 2-13/16 x 3-5/8	16	3.75	370	22.00	8.0
.17	Saxon 4"	11-5/8 x 3-5/8 x 3-5/8	19	3.00	320	26.00	9.0
	6"	11-5/8 x 3-5/8 x 5-5/8	28	3.00	260	35.00	13.5
	8"	11-5/8 x 3-5/8 x 7-5/8	38	3.00	200	50.00	18.0

Deduct from above dimensions for 1/2" Joint 1/8" x 1/8" x 1/8".
Add to above dimensions for full size 3/8" x 0" x 2/8".

		Size - Inches (l x h x d)	Mortar CuFt M	Unit Per SqFt	Units Placed ManDay	Haul & Unload M PCs	Unit Wt.
.2	Coated Brick (Glazed)		12	6.75	510	17.00	4.5
.3	Common Brick						
	Clay	7-14 x 2-1/16 x 3-1/2	13	6.75	660	15.00	4.5
	Concrete		13	6.75	660	15.00	4.5
.4	Fire Brick	8-0 x 2- 1/4 x 3-1/2	13	6.65	660	15.00	7.0
0402.0	**STANDARD AND LIGHT-WEIGHT CONCRETE BLOCK**						
	12" Plain	11-5/8 x 7-5/8 x 15-5/8	65	1.125	190	110.00	5.0
	1/2 Height	11-5/8 x 3-5/8 x 15-5/8		2.25		Lt Wt	38.0
	1/2 Length	11-5/8 x 7-5/8 x 7-5/8		2.25			
	8" Plain	7-5/8 x 7-5/8 x 15-5/8	55	1.125	200	80.00	36.0
	1/2 Height	7-5/8 x 3-5/8 x 15-5/8		2.25		Lt Wt	27.0
	1/2 Length	7-5/8 x 7-5/8 x 7-5/8		2.25			
	6" Plain	5-5/8 x 7-5/8 x 15-5/8	45	1.125	210	60.00	28.0
	1/2 Height	5-5/8 x 3-5/8 x 15-5/8		2.25		Lt Wt	21.0
	1/2 Length	5-5/8 x 7-5/8 x 7-5/8		2.25			
	4" Plain	3-5/8 x 7-5/8 x 15-5/8	35	1.125	210	50.00	24.0
	1/2 Height	3-5/8 x 3-5/8 x 15-5/8		2.25		Lt Wt	18.0
	1/2 Length	3-5/8 x 7-5/8 x 7-5/8		2.25			
	16" Plain	15-5/8 x 7-5/8 x 15-5/8	70	2.25	170	150.00	61.0
	14" Plain	13-5/8 x 7-5/8 x 15-5/8	65	2.25	180	90.00	56.0
	10" Plain	9-5/8 x 7-5/8 x 15-5/8	60	2.25	195	85.00	43.0

Average 2 cubic yards mortar per M Block.

TABLE B

		Modular Manufactured Size - Inches l		h		d	Mortar CuFt M	Unit Per SqFt	Units Placed ManDay	Haul & Unload M PCs	Unit Wt.
0403.0	**CLAY BACKING & PARTITION TILE**										
	3"	3	x	11-1/2	x	11-1/2	18.7	1.00	260	25.0	13.0
	4"	4	x	11-1/2	x	11-1/2	22.2	1.00	250	29.0	15.0
	6"	6	x	11-1/2	x	11-1/2	16.0	1.00	230	38.0	20.0
0404.0	**CLAY FACING TILE (GLAZED)**										
	2" 6T	1-3/4	x	5-1/16	x	7-3/4	13	2.25	180	20.00	6.5
	4" 6T	3-3/4	x	5-1/16	x	11-3/4		2.25		30.00	10.5
	6" 6T	5-3/4	x	5-1/16	x	11-3/4		2.25		45.00	15.2
	8" 6T	7-3/4	x	5-1/16	x	11-3/4		2.25		55.00	19.0
	2" 8W	1-3/4	x	7-3/4	x	15-3/4	22	1.13	125	40.00	14.0
	4" 8W	3-3/4	x	7-3/4	x	15-3/4		1.13		60.00	21.0
	6" 8W	5-3/4	x	7-3/4	x	15-3/4		1.13		85.00	31.0
0405.0	**GLASS UNITS**										
	6" x 6"	5-3/4	x	5-3/4	x	3-7/8	12.5	4.00	140	18.00	4.0
	6" x 8"	7-3/4	x	7-3/4	x	3-7/8	16.0	2.25	125	26.00	7.0
	12" x 12"	11-3/4	x	11-3/4	x	3-7/8	23.3	1.00	90	33.00	16.0
0406.0	**TERRA-COTTA**										
	3"	3"	x	12"	x	12"	25	1.00		36.00	10.0
	4"	4"	x	12"	x	12"	30	1.00		40.00	13.0
	6"	6"	x	12"	x	12"	40	1.00		47.00	18.0
	8"	8"	x	12"	x	12"	60				
0407.0	**MISCELLANEOUS UNITS**										
	Coping Tile	9"					50	-	160	35.00	-
		12"					60	-	140	40.00	-
	Flue Lining	8"	x	8"			50	-	160	35.00	-
		12"	x	12"			55	-	130	45.00	-
		18"	x	18"			65	-	60	55.00	-
		21"	x	21"			75	-	40	70.00	-
		24"	x	24"			90	-	30	90.00	-
0408.0	**NATURAL STONE**										
	Cut Stone						SqFt				CuFt
	Lime Stone						10.0	-	18	25.00	150.0
	Marble						10.0	-	13	25.00	160.0
	Granite						10.0	-	11	25.00	165.0
	Ashlar Stone						15.0	-	26	25.00	150.0
	Rough Stone - Flagstone						15.0	-	22	25.00	140.0
	Field Stone						15.0	-	22	25.00	160.0
	Light Weight Stone						15.0	-	24	25.00	220.0
0409.0	**PRECAST VENEERS**										
	Architectural Precast Stone						10.0	-	18	25.00	150.0
	Terra-Cotta						10.0	-	17	25.00	130.0
	Precast Concrete						10.0	-	24	25.00	150.0

		UNIT	LABOR	MATERIAL
0501.0	**STRUCTURAL METAL (L+M) (Ironworkers)**			
.1	BEAMS AND COLUMNS			
.11	Wide Flange Shapes, H-Bearing Piles			
	Light Weight Flange Columns & Misc. Columns		(Subcontracted)	
	Light Beams - American Standard			
	Small Job (5 to 30 ton)	Ton	460.00	1,450.00
	Medium Job (30 to 100 ton)	Ton	450.00	1,420.00
	Large Job (100 ton and up)	Ton	420.00	1,380.00
.12	Junior Beams	Ton	435.00	1,480.00
.2	CHANNELS - American Std. - Junior -3" to 15"	Ton	590.00	1,300.00
	10" to 12"	Ton	570.00	1,460.00
.3	ANGLES (Equal and Unequal)	Ton	640.00	1,600.00
.4	STRUCTURAL T's (from wide Flange Shapes)			
	Light Shapes and American Standard	Ton	460.00	1,400.00
.5	STRUCTURAL TUBING (Steel Pipe, Square & Rectangular)	Ton	470.00	1,430.00
.6	PLATES AND BARS	Ton	490.00	1,500.00
.7	BOX GIRDERS	Ton	790.00	1,440.00
.8	TRUSSES	Ton	830.00	1,950.00
	Add for Welded Field Connections	Ton	185.00	-
	Add for High Strength Bolting	Ton	-	185.00
	Add for Riveted	Ton	350.00	-
	Add for Weathering Steel	Ton	-	260.00
	Add for Galvanizing	Ton	-	375.00
0502.0	**MISCELLANEOUS METALS (L+M) (Ironworkers)**			
.1	BOLTS (with Nuts & Washers in Concrete or Masonry) See 0201.36 for Hole Drilling in Field			
	1/2" x 4"	Each	2.20	1.75
	5/8" x 4"	Each	2.30	1.95
	3/4" x 4"	Each	2.50	2.50
	1" x 4"	Each	2.70	3.25
.2	CASTINGS			
	Manhole Ring & Cover -24" Light Duty	Each	82.00	280.00
	24" Heavy Duty	Each	92.00	340.00
	Wheel Guards- 2' 6"	Each	86.00	215.00
.3	CORNER GUARDS- 3" x 3" x 1/4" Steel	LnFt	5.40	6.70
	4" x 4" x 1/4" Steel	LnFt	6.00	8.50
	Galv.	LnFt	5.50	10.40
.4	FIRE ESCAPES- 24" Steel Grate	Per Floor	1,030.00	3,000.00
		Tread	64.00	180.00
.5	FRAMES, CHANNEL- 6" - 8.2 lbs	LnFt	4.60	11.30
	8" - 11.5 lbs	LnFt	6.40	15.40
	10" - 15.3 lbs	LnFt	8.75	27.00
.6	GRATINGS- Steel - 1" x 1/8"	SqFt	2.30	8.00
	Aluminum - 1" x 1.8"	SqFt	2.30	16.50
	Cast Iron #2 - Light Duty	SqFt	2.30	11.00
	Heavy Duty	SqFt	2.00	21.00
.7	GRILLES	SqFt	3.50	9.50
.8	HANDRAILS - 1 1/2"			
	Wall Mounted - Single Rail- Steel	LnFt	6.40	11.10
	Aluminum	LnFt	7.50	41.00
	Floor Mounted - 2 Rail- Steel	LnFt	8.65	24.50
	3 Rail - Steel	LnFt	9.60	30.20
	Add for Galvanizing	LnFt	-	20%
.9	LADDERS - Steel	LnFt	11.75	38.00
	Aluminum	LnFt	11.30	5.00
.10	LINTELS, LOOSE	Ton	540.00	1,120.00
	3" x 3" x 1/4" - 4.9 lbs per foot	LnFt	1.80	4.20
	4" x 3" x 1/4" - 5.8 lbs per foot	LnFt	2.20	4.70
	4" x 3 1/2" x 1/4" - 6.2 lbs per foot	LnFt	2.35	5.00
	4" x X3 1/2" x 5/16" - 7.7 lbs per foot	LnFt	3.00	5.80
	4" x X3 1/2" x 3/8" - 9.1 lbs per foot	LnFt	3.45	6.80
	4" x 4" x 1/4" - 6.6 lbs per foot	LnFt	2.80	5.60
	Add for Punching Holes	Each	-	3.90
	Add for Galvanizing	LnFt	-	1.20

		UNIT	LABOR	MATERIAL (Subcontracted)
0502.0	**MISCELLANEOUS METALS (Cont'd...)**			
.11	PIPE GUARDS - 4" - 4' Above Grade &	LnFt	10.05	8.60
	(Bollards in Concrete) - 4' Below Grade	LnFt	10.60	11.00
.12	PLATE, CHECKERED - 1/4" - 10.2 lbs	SqFt	1.80	9.10
	1/8" - 6.2 lbs	SqFt	1.45	8.15
.13	STAIRS, METAL PAN - 4' Wide x 10' (No Concrete)	PerFl	900.00	4,100.00
	Including Supports	Tread	63.00	275.00
.14	STAIRS, CIRCULAR - 3' x 10' - Aluminum	PerFl	505.00	4,200.00
	Steel	PerFl	555.00	4,400.00
	Cast Iron	PerFl	610.00	4,770.00
.15	WIRE GUARDS	SqFt	2.40	10.80
.16	TRENCH FRAME AND COVER - 2'	LnFt	6.25	11.50
0503.0	**ORNAMENTAL METALS (M) (Ironworkers)**			
	Non Ferrous - See 0502.0			
0503.0	**OPEN WEB JOISTS (L+M) (Ironworkers)**			
.1	STANDARD BAR JOISTS (Spans to 48')			
	Small Jobs (to 20 ton) - Spans to 15'	Ton	360.00	1,120.00
	25'	Ton	365.00	1,100.00
	35'	Ton	350.00	1,070.00
	Medium Jobs (20 - 50 ton) - Spans to 15'	Ton	350.00	1,120.00
	25'	Ton	345.00	1,090.00
	35'	Ton	340.00	1,070.00
	48'	Ton	330.00	1,040.00
	Large Jobs (50 ton & up) - Spans to 15'	Ton	340.00	1,110.00
	25'	Ton	330.00	1,090.00
	35'	Ton	330.00	1,070.00
	48'	Ton	320.00	1,060.00
.2	LONG SPAN JOISTS (Spans 48' to 150')			
	Small Jobs (to 30 ton) - Spans to 60'	Ton	360.00	1,110.00
	Medium Jobs (30 -100 ton) - Spans to 60'	Ton	355.00	1,120.00
	80'	Ton	340.00	1,145.00
	100'	Ton	320.00	1,155.00
	Large Jobs (100 ton & Up) - Spans to 60'	Ton	335.00	1,100.00
	80'	Ton	330.00	1,140.00
	100'	Ton	325.00	1,170.00
	120'	Ton	315.00	1,240.00
	150'	Ton	310.00	1,290.00
0505.0	**METAL DECKING (L+M) (Sheet Metal Workers)**			
.1	SHORT SPAN (Spans to 6')			
	(Baked Enamel - Ribbed, V Beam & Seam)			
	1 1/2" Deep - 18 Ga	Sq	54.00	160.00
	20 Ga	Sq	50.00	125.00
	22 Ga	Sq	49.00	120.00
	Add for Galvanized	Sq	-	10%
.2	LONG SPAN (Spans to 12') (Enameled & Open Bottom)			
	3" Deep - 16 Ga	Sq	70.00	490.00
	18 Ga	Sq	65.00	440.00
	20 Ga	Sq	66.00	390.00
	22 Ga	Sq	63.20	320.00
	4 1/2" Deep - 16 Ga	Sq	75.10	580.00
	18 Ga	Sq	74.20	540.00
	20 Ga	Sq	72.80	490.00
	Add for Acoustical	Sq	-	20%
	Add for Galvanized	Sq	-	20%

			UNIT	LABOR (Subcontracted)	MATERIAL
0505.0	**METAL DECKING (L+M) (Sheet Metal), Cont'd...**				
.3	CELLULAR - Electric and Air Flow				
	3" Deep - 16 - 16 Ga Galvanized		Sq	84.60	660.00
	18 - 18 Ga Galvanized		Sq	78.70	630.00
	4 1/2" Deep- 16 - 16 Ga Galvanized		Sq	98.50	880.00
	18 - 18 Ga Galvanized		Sq	92.20	740.00
.4	CORRUGATED METAL				
	Standard	Spans to Approx.			
	Black - .015"	3' 6"	Sq	40.60	75.00
	Galvanized - .015"	3' 6"	Sq	40.60	90.00
	Heavy Duty				
	Black - 26 Ga	5' 0"	Sq	43.70	88.00
	Galvanized - 26 Ga	5' 0"	Sq	43.70	93.00
	Super Duty				
	Black - 24 Ga	7' 0"	Sq	45.80	127.00
	Galvanized - 24 Ga	7' 0"	Sq	45.80	136.00
	Black - 22 Ga	7' 6"	Sq	46.85	146.00
	Galvanized - 22 Ga	7' 6"	Sq	46.85	150.00
.5	BOX RIB				
	2" - 24 Ga		Sq	63.30	445.00
	22 Ga		Sq	67.60	470.00
	20 Ga		Sq	71.80	500.00
0506.0	**METAL SIDINGS & ROOFINGS (L+M) (Sheet Metal)**				
.1	INDUSTRIAL				
.11	Steel (24 Ga - See 0505 for Other Gauges)				
	Corrugated - Galvanized		Sq	78.60	132.00
	Ribbed - Enameled		Sq	72.20	128.00
	Add for Liner Panel		Sq	35.90	120.00
	Add for Insulation		Sq	30.65	32.00
.12	Aluminum- Corrugated		Sq	68.00	120.00
.2	ARCHITECTURAL (Prefinished)				
.21	Steel - Baked Enameled		Sq	103.70	215.00
	Porcelain Enameled		Sq	131.00	265.00
	Acrylic Enameled		Sq	125.00	230.00
	Plastic Faced		Sq	125.00	290.00
.22	Aluminum - Anodized		Sq	122.00	215.00
	Plastic Faced		Sq	124.00	290.00
	Porcelainized		Sq	128.00	280.00
	Add for Liner Panel and Insulation		Sq	48.00	125.00
.23	Stainless Steel		Sq	202.00	650.00
.24	Protected Metal (Asphalt on Steel with Finish)		Sq	148.00	450.00
	See 1310 for Metal Buildings				
0507.0	**LIGHT GAGE FRAMING (L+M) (Ironworkers)**				
	3 5/8"- 16 Ga		LnFt	.65	3.90
	6"- 16 Ga		LnFt	.68	4.85
	8"- 16 Ga		LnFt	.82	5.60
	10"- 16 Ga		LnFt	.98	7.00
	Average - Wall		Lb	.38	1.00
	Average - Hanging		Lb	.77	1.20
0508.0	**EXPANSION CONTROL (L+M) (Ironworkers)**				
	4"- Aluminum		LnFt	7.60	40.50
	Bronze		LnFt	8.05	112.00
	Stainless Steel		LnFt	8.05	115.00
	2"- Aluminum		LnFt	6.10	36.50
	Bronze		LnFt	7.55	66.40
	Stainless Steel		LnFt	7.55	73.50

		Sq. Ft. Cost
0501.0	**Structural Steel Frame**	
	To 30 Ton 20' Span	7.35
	24' Span	7.85
	28' Span	8.24
	32' Span	9.35
	36' Span	10.40
	40' Span	11.25
	44' Span	11.90
	48' Span	12.80
	52' Span	14.45
	56' Span	15.90
	60' Span	16.85
	Deduct for Over 30 Ton	5%
0504.0	**Open Web Joists**	
	To 20 Ton 20' Span	2.40
	24' Span	2.45
	28' Span	2.50
	32' Span	2.65
	36' Span	2.85
	40' Span	3.30
	48' Span	4.10
	52' Span	5.30
	56' Span	4.90
	60' Span	5.40
	Deduct for Over 20 Ton	10%
0505.0	**Metal Decking**	
	1 1/2" Deep - Ribbed - Baked Enamel - 18 Ga	3.00
	20 Ga	2.90
	22 Ga	2.50
	3" Deep - Ribbed - Baked Enamel - 18 Ga	6.05
	20 Ga	5.70
	22 Ga	5.15
	4 1/2" Deep - Ribbed - Baked Enamel - 16 Ga	7.90
	18 Ga	6.80
	20 Ga	6.30
	3" Deep - Cellular - 18 Ga	7.70
	16 Ga	9.65
	4 1/2" Deep - Cellular - 18 Ga	10.80
	16 Ga	12.30
	Add for Galvanized	10%
	Corrugated Black Standard .015	1.50
	Heavy Duty - 26 Ga	1.60
	S. Duty - 24 Ga	2.15
	22 Ga	2.20
	Add for Galvanized	15%
0506.0	**Metal Sidings and Roofings**	
	Aluminum- Anodized	4.10
	Porcelainized	8.25
	Corrugated	2.70
	Enamel, Baked Ribbed	4.25
	24 Ga Ribbed	2.90
	Acrylic	5.05
	Porcelain Ribbed	5.60
	Galvanized- Corrugated	2.75
	Plastic Faced	6.00
	Protected Metal	7.30
	Add for Liner Panels	2.25
	Add for Insulation	.75

0601.0 ROUGH CARPENTRY (Carpenters)

		UNIT	LABOR	MATERIAL
.1	LIGHT FRAMING & SHEATHING (Standard - Hem-Fir)			
.11	Joists, Beams and Headers - 16" O.C. to 14'			
	2"x 6" Floor	BdFt	.61	.51
	2"x 8" Floor	BdFt	.59	.54
	2"x10" Floor	BdFt	.57	.56
	2"x12" Floor	BdFt	.59	.58
	Add for Ceiling Joists - 2nd Fl or Roof	BdFt	.12	-
	Add for Door, Window & Misc. Headers	BdFt	.40	-
	Add for Sloped Installations	BdFt	40%	-
	Add for Ledgers	BdFt	.25	-
.12	Studs, Plates and Miscellaneous Framing - 16" O.C.			
	2" x 2"	BdFt	.85	.54
	2" x 3"	BdFt	.80	.52
	2" x 4"	BdFt	.69	.50
	2" x 6"	BdFt	.67	.55
	Add for 2nd Floor and Above per Floor	BdFt	.12	-
	Add for Bolted Plates or Sills	BdFt	.60	-
	Add for Each 1' Above 8' in Length	BdFt	.04	-
	Add for Fire Stops, Fillers, and Nailers	BdFt	.38	-
	Add for Soffit & Suspended Framing 2"x 4"	BdFt	.85	-
	See 0602 for Metal Studs and Plates			
.13	Bridging - 16" O.C.			
	1" x 3" Wood Diagonal	Set	1.52	.40
	2" x 4" Diagonal	Set	1.72	.95
	1" Metal Diagonal - 18 Ga	Set	1.52	.95
	2" x 8" Solid	BdFt	1.78	.68
.14	Rafters and Field-Made Trussed Rafters - 16" O.C.			
	(See 0605 for Prefabrication)			
	2" x 4"	BdFt	.90	.59
	2" x 6"	BdFt	.88	.57
	2" x 8"	BdFt	.85	.54
	2" x 10"	BdFt	.80	.58
	2" x 4" Trusses - Make up in Field	BdFt	.70	.74
	Erection	BdFt	.48	-
	Add for Hips and Valleys	BdFt	.48	-
	Add for Bracing	BdFt	.90	.62
.15	Stairs			
	2" x 10" and 2" x 12" for Stringers	BdFt	1.95	.58
	1" x 8" and 2" x 10" for Risers & Stairs	BdFt	1.38	.59
	Average for Stringers, Treads and Risers	BdFt	1.65	.72
	Add to Above Framing Where Applicable:			
	Circular Installations	BdFt	60%	-
	Diagonal Installations	BdFt	10%	7%
	Kiln Dry Lumber	BdFt	-	15%
	Construction Grade Lumber	BdFt	-	.06
	Lumber Over 14'	BdFt	-	.05
	Winter Construction	BdFt	10%	-
	Add for Pressure Treating			
	2" x 4" - 2" x 6" - 2" x 8"	BdFt	-	.16
	2" x 10" - 2" x 12"	BdFt	-	.25
	Add for Fire Treating - .40	BdFt	-	.16
	- Dricon	BdFt	-	.30

0601.0 ROUGH CARPENTRY, Cont'd...

		UNIT	LABOR	MATERIAL
.16	Sub Floor Sheathing (Structural)			
	1" x 8" and 1" x 10" #3 Pine	BdFt	.42	.70
	1/2" x 4' x 8' Plywood - CD Exterior	SqFt	.40	.57
	5/8" x 4' x 8' Plywood - CD Exterior	SqFt	.41	.67
	3/4" x 4' x 8' Plywood - CD Exterior	SqFt	.42	.77
	3/4" x 4' x 8' Plywood - T&G, CD Exterior	SqFt	.48	1.00
	1/2" x 4' x 8' OSB Board	SqFt	.42	.75
	5/8" x 4' x 8' OSB Board	SqFt	.43	.90
	3/4" x 4' x 8' OSB Board, T&G	SqFt	.47	1.10
.17	Floor Sheathing (Over Sub Floor) With Partitions in Place			
	1/4" x 4' x 8' Plywood - CD	SqFt	.43	.43
	3/8" x 4' x 8' Plywood - CD	SqFt	.44	.44
	1/2" x 4' x 8' Plywood - CD	SqFt	.45	.57
	5/8" x 4' x 8' Plywood - CD	SqFt	.46	.67
	3/8" x 4' x 8' Particle Board	SqFt	.44	.26
	1/2" x 4' x 8' Particle Board	SqFt	.45	.31
	5/8" x 4' x 8' Particle Board	SqFt	.46	.39
	3/4" x 4' x 8' Particle Board	SqFt	.47	.45
.18	Wall Sheathing			
	1" x 8" and 1" x 10" #3 Pine	BdFt	.47	.70
	3/8" x 4' x 8' Plywood - CD Exterior	SqFt	.46	.46
	1/2" x 4' x 8' Plywood - CD Exterior	SqFt	.47	.57
	5/8" x 4' x 8' Plywood - CD Exterior	SqFt	.48	.67
	3/4" x 4' x 8' Plywood - CD Exterior	SqFt	.50	.78
	1/2" x 4' x 8' Fiberboard Impregnated	SqFt	.45	.27
	25/32" x 4' x 8' Fiberboard Impregnated	SqFt	.47	.30
	1" x 2' x 8' T&G Styrofoam	SqFt	.48	.34
	2" x 2' x 8' T&G Styrofoam	SqFt	.64	.65
	1/2" x 4' x 8' OSB Board	SqFt	.46	.75
	5/8" x 4' x 8' OSB Board	SqFt	.49	.90
	3/8" x 4' x 8			1.10
.19	Roof Sheathing - Flat			
	1" x 6" and 1" x 8" #3 Pine	BdFt	.47	.70
	1/2" x 4' x 8' Plywood - CD Exterior	SqFt	.44	.57
	5/8" x 4' x 8' Plywood - CD Exterior	SqFt	.46	.67
	3/4" x 4' x 8' Plywood - CD Exterior	SqFt	.47	.77
	Add for Sloped Installation	SqFt	.14	-
	Add Steep Sloped Installation (over 5-12)	SqFt	.30	-
	Add for Dormer Work	SqFt	.60	-
	Add to Above Sheathing where Applicable:			
	Fire Treating	SqFt	-	.32
	Pressure Treating	SqFt	-	.22
	Winter Construction	SqFt	10%	-
	Plywood - AC Exterior (Good One Side)	SqFt	-	.28
	Plywood - AD Exterior (Good One Side)	SqFt	-	.34
	Diagonal Installations	SqFt	8%	7%
	Plywood 10' Lengths	SqFt	-	.23

Recommended 10% Decrease on Labor Units for Residential Construction

0601.0 ROUGH CARPENTRY, Cont'd...

		UNIT	LABOR	MATERIAL
.2	HEAVY FRAMING AND DECKING (Construction-Grade Fir, Air Dried S4S) See 0604.0 for Laminated Heavy Framing & Decking. See 0606.0 for Wood Treatments. See 0109.8 for Wood Curbs and Walls.			
.21	Columns and Posts			
	4"x 4"	MBF	580.00	850.00
	4"x 6"	MBF	575.00	870.00
	6"x 6"	MBF	530.00	1,030.00
	8"x 8"	MBF	525.00	1,150.00
	12"x12"	MBF	500.00	1,600.00
	Add for Fitting at Base and Heat	MBF	170.00	-
	Beams and Joists			
	4" x 4"	MBF	580.00	850.00
	4" x 6"	MBF	575.00	870.00
	4" x 8"	MBF	570.00	990.00
	4" x10"	MBF	535.00	1,050.00
	6" x 6"	MBF	540.00	1,040.00
	6" x 8"	MBF	525.00	1,080.00
	6" x10"	MBF	520.00	1,140.00
	8" x12"	MBF	510.00	1,340.00
	Add to 0601.21 Above for:			
	Rough Sawn	MBF	-	130.00
	Mill Cutting to Length and Shape	MBF	-	130.00
	Nails	MBF	-	65.00
	Hauling & Unloading (if Quoted Cars)	MBF	-	85.00
	Equipment & Operator Cost - Fork Lift	MBF	-	70.00
	Equipment & Operator Cost - Crane	MBF	-	100.00
	Add for Red Cedar	MBF	-	1,150.00
	Add for Redwood	MBF	-	1,360.00
.22	DECKING			
	Tongue & Groove			
	Fir, Hemlock & Spruce, Construction Grade (SPF)			
	4" x 4"	MBF	370.00	930.00
	2" x 6" and 2" x 8"	MBF	345.00	700.00
	3" x 6"	MBF	360.00	850.00
	4" x 6"	MBF	370.00	960.00
	Cedar - #3 and better			
	3" x 4"	MBF	355.00	1,540.00
	4" x 4" - 4" x 6" - 4" x 8" and 4" x 10"	MBF	350.00	1,680.00
	White Pine			
	3" x 6" and 3" x 8"	MBF	355.00	1,600.00
	4" x 6" and 4" x 8"	MBF	365.00	1,730.00
	Add for D Grade	MBF	-	1,100.00
	2" x 6" Panelized Decking - 1 1/2" x 20"			
	Premium Grade	SqFt	.76	2.25
	Architectural Grade	SqFt	.76	1.90
	Add for Select Grade	MBF	-	5%
	Add for Hemlock	MBF	-	50.00
	Add for Hip Roof	MBF	70.00	-
	Add for Specified Lengths	MBF	-	35.00
	Add for End Matched	MBF	-	44.00
	Add for Hauling and Unloading	MBF	-	85.00
	Add for Nails	MBF	-	55.00
	See 0604 for Laminated Decking			
	See 0604.3 for Square Foot Deck Costs			

0601.0 ROUGH CARPENTRY, Cont'd...

		UNIT	LABOR	MATERIAL
.3	MISCELLANEOUS CARPENTRY			
.31	Blocking and Bucks (2" x 4" or 2" x 6")			
	Doors and Window Bucks			
	Erected Before Masonry	BdFt	1.45	.52
		or, Each	25.00	8.80
	Bolted to Masonry or Steel	BdFt	2.00	.52
		or, Each	34.00	8.80
	Nail Driven to Steel Studs or Masonry	BdFt	1.30	.52
		or, Each	22.00	8.80
	Roof Blocking - Pressure Treated			
	Edges - Bolted to Concrete	BdFt	1.52	.80
	Nailed to Wood	BdFt	.76	.75
	Openings - Nailed	BdFt	1.70	.75
	Add for 2" x 8" or 2" x 10"	BdFt	-	.05
	Add to 0601.31 for Bolts - 4' O.C.	BdFt	-	.30
.32	Grounds, Furring, Sleepers & Nailers #3 Pine			
	1" x 8" Fastened to Wood	LnFt	.55	.47
	1" x 6" Fastened to Wood	LnFt	.52	.34
	1" x 4" Fastened to Wood	LnFt	.49	.22
	1" x 3" Fastened to Wood	LnFt	.46	.19
	1" x 3" Fastened to Concrete/ Masonry			
	Nailed	LnFt	.62	.30
	Gun Driven (Incl Fasteners)	LnFt	.55	.35
	Pre Clipped	LnFt	.80	.32
	Add for Ceiling Work	LnFt	.18	-
	2" x 4" Not Suspended Framing	BdFt	1.50	.52
	Suspended Framing	BdFt	1.85	.52
	2" x 2" Not Suspended Framing	BdFt	1.90	.56
	Suspended Framing	BdFt	2.70	.56
.33	Cant Strips			
	Pressure Treated			
	Cut from 4" x 4" - At Roof Edges	BdFt	.47	.43
		or, LnFt	.63	.64
	4" x 4" - At Roof Openings	BdFt	.68	.43
		or, LnFt	.91	.64
	6" x 6" - At Roof Edges	BdFt	1.65	2.20
		or, LnFt	1.10	1.45
	6" x 6" - At Roof Openings	BdFt	2.28	2.20
		or, LnFt	1.52	1.45
	Add for Bolting (Bolts - 4' O.C.)	BdFt	.68	.55
	Not Treated			
	Cut from 4" x 4" - At Roof Edges	BdFt	.46	.35
		or, LnFt	.62	.52
	6" x 6" - At Roof Edges	BdFt	1.65	1.65
		or, LnFt	1.10	1.10
.34	Building Papers and Sealers			
	15" Felt (432 Sq.Ft. to Roll)	SqFt	.07	.03
	20" Felt (432 Sq.Ft. to Roll)	SqFt	.08	.04
	30" Felt (216 Sq.Ft. to Roll)	SqFt	.09	.06
	Cotton and Glass Asphalt Saturated	SqFt	.08	.06
	Polyethylene 4 mil	SqFt	.07	.02
	6 mil	SqFt	.08	.03
	30# Red Rosin Paper	SqFt	.08	.13
	Sill Sealer 6" x 100'	SqFt	.25	.21

0602.0 FINISH CARPENTRY (Carpenters)

		UNIT	LABOR	MATERIAL
.1	FINISH SIDINGS AND FACING MATERIAL (Exterior) (See Quick Estimating for Square Foot Costs)			
.11	Boards, Beveled and Lap Siding			
	Cedar - Beveled - Clear 1/2" x 4"	BdFt	1.10	1.60
	1/2" x 6"	BdFt	1.03	1.72
	Beveled - Rough Sawn 7/8" x 8"	BdFt	.95	1.48
	7/8" x 10"	BdFt	.91	1.70
	7/8" x 12"	BdFt	.89	1.75
	Board- Rough Sawn 3/4" x 12"	BdFt	.87	1.70
	Redwood Beveled- Clear Heart			
	1/2" x 6"	BdFt	1.02	1.95
	1/2" x 8"	BdFt	.95	2.05
	5/8" x 10"	BdFt	.90	2.75
	3/4" x 6"	BdFt	.93	2.55
	3/4" x 8"	BdFt	.89	2.65
	Board - Clear Heart 3/4" x 6"	BdFt	.86	3.05
	3/4" x 8"	BdFt	.83	4.15
	Board, - T&G 1" x 4" & 6"	BdFt	.89	3.50
	Rustic Beveled - 1" x 8" & 10"	BdFt	.86	3.70
	5/4"x 6" & 8"	BdFt	.83	3.60
	Add for Metal Corners	Each	.90	.35
	Add for Mitering Corners	Each	2.90	-
.12	Plywood - Siding			
	Fir - AC - Smooth One Side 1/4"	SqFt	.58	.81
	3/8"	SqFt	.60	.92
	Fir - Rough Faced 3/8"	SqFt	.62	.98
	Fir - AC - Smooth One Side 1/2"	SqFt	.63	.91
	5/8"	SqFt	.66	1.01
	3/4"	SqFt	.70	1.11
	Fir - Grooved and Rough Faced 5/8"	SqFt	.68	1.10
	Cedar - Rough Sawn 3/8"	SqFt	.61	1.10
	5/8"	SqFt	.68	1.35
	3/4"	SqFt	.71	1.52
	Cedar - Grooved & Rough Faced (4" & 8") 5/8"	SqFt	.68	1.55
	Add for 4' x 10' and 4' x 9' Sheets	SqFt	-	.15
	Add for Wood Batten Strips 1" x 2" Pine	LnFt	.60	.30
	Add for Horizontal Joint Flashing (Spline)	LnFt	.48	.40
	Add for Medium Density Fir	SqFt	.02	.16
.13	Hardboard - Paneling and Lap Siding (Primed)			
	7/16" - Rough Textured 4' x 8' Paneling	SqFt	.52	.98
	7/16" - Grooved 4' x 8' Paneling	SqFt	.58	.98
	7/16" - Stucco Board 4' x 8' Paneling	SqFt	.59	1.05
	Lap Siding - 7/16" x 8"	BdFt	.79	.70
	7/16"x 12"	BdFt	.86	.75
	Add for Prefinishing	BdFt	-	.16
.14	Shingles - Wood			
	16" Red Cedar - 12" to Weather - #1	SqFt	.82	1.90
	#2	SqFt	.79	1.50
	#3	SqFt	.71	1.15
	Fancy Butt Red Cedar 7" #1	SqFt	.84	3.90
	Red Cedar Hand Split - #1 1/2" to 3/4"	SqFt	.98	1.50
	3/4" to 1 1/4" - #1	SqFt	1.02	1.80
	3/4" to 1 1/4" - #2	SqFt	.99	1.90
	3/4" to 1 1/4" - #3	SqFt	.96	1.80
	Add for Fire Retardant	SqFt	-	.42
	Add for 3/8" Backer Board	SqFt	.30	.24
	Add for Metal Corners	Each	.82	.35
	Add for Ridges, Hips and Corners	LnFt	1.38	.85
	Add for 8" to Weather	SqFt	25%	-

See Division 0708 for Roof Shingles
No Finish Included in Above Costs

0602.0 FINISH CARPENTRY, Cont'd...

		UNIT	LABOR	MATERIAL Pine #2	MATERIAL Cedar #3	MATERIAL Plywood 5/8" ACX	
.15	Facias and Trim						
	All Sizes (average)	BdFt	.80	.78	1.63	.90	
	1" x 3"	LnFt	.50	.19	.41	.22	
	1" x 4"	LnFt	.58	.22	.50	.30	
	1" x 6"	LnFt	.68	.34	.87	.45	
	1" x 8"	LnFt	.73	.47	1.20	.60	
	1" x 12"	LnFt	.80	.77	1.63	.90	
	Add for Clear Grades			400%	100%	-	30%

		UNIT	LABOR	MATERIAL
.16	Soffits			
	1/2" Plywood Fir ACV	SqFt	1.00	.65
	Aluminum	SqFt	1.00	1.25
	Plastic - Egg Crate	SqFt	1.00	1.30
.2	FINISH WALLS AND CEILING MATERIALS (Interior)			
.21	Boards (See Quick Est. - Size/Cutting Allowance Added)			
	Cedar - #3 Grade - 1" x 6"	SqFt	.83	1.90
	Knotty 1" x 6"	SqFt	.81	2.00
	1" x 8"	SqFt	.78	2.20
	D Grade 1" x 4"	SqFt	.83	2.80
	1" x 6"	SqFt	.81	2.90
	1" x 8"	SqFt	.79	2.80
	Aromatic 1" x 6"	SqFt	.83	2.90
	Redwood - Constr.1" x 6"	SqFt	.87	2.15
	1" x 8"	SqFt	.81	2.35
	Add for Clear	SqFt	10%	100%
	Fir - Beaded 5/8" x 4"	SqFt	.87	1.95
	#2 Grade 1" x 6"	SqFt	.82	2.15
	1" x 8"	SqFt	.80	2.10
	1" x 10"	SqFt	.74	2.15
.22	Hardboard & Pressed Woods (4' x 8' Panels)			
	Tempered 1/8"	SqFt	.53	.30
	1/4"	SqFt	.59	.39
	Pegboard (Perforated)1/8"	SqFt	.55	.36
	1/4"	SqFt	.60	.49
	Plastic Faced 1/4"	SqFt	.63	.60
	Flake Board 1/8"	SqFt	.49	.30
	Add for Metal or Plastic Mouldings	LnFt	.51	.40
	Add for Pre-finished	SqFt	-	.18
.23	Plywood Paneling (4'x 8' Panels, Prefinished)			
	Birch, Natural 1/4"	SqFt	.85	1.00
	3/4"	SqFt	.96	1.65
	White Select 1/4"	SqFt	.86	1.60
	Oak, Red - Rotary Cut 1/4"	SqFt	.86	.90
	3/4"	SqFt	.98	2.00
	White 1/4"	SqFt	.85	2.00
	Cedar, Aromatic 1/4"	SqFt	.86	1.65
	Cherry 3/4"	SqFt	.98	2.30
	Pecan 3/4"	SqFt	.98	2.30
	Walnut 1/4"	SqFt	.85	2.30
	3/4"	SqFt	.98	3.60
	Mahogany, Lauan 1/4"	SqFt	.85	.70
	3/4"	SqFt	.98	1.20
	African 3/4"	SqFt	.98	2.50
	Knotty Pine 1/4"	SqFt	.85	1.15
	3/4"	SqFt	.98	1.90
	Slat Wall - Painted 3/4"	SqFt	.98	1.50
	Oak 3/4"	SqFt	1.05	2.35
	Add for V Groove and T&G	SqFt	-	.27
	Add for Glued-On Applications	SqFt	.17	.10
	Add for 10' Lengths	SqFt	.12	.25

0602.0 FINISH CARPENTRY, Cont'd...

		UNIT	LABOR	MATERIAL
.24	Gypsum Board and Framing			
	(See 0902 - Subcontracted Work Costs)			
	(See 0902 - Metal Framing SqFt Costs)			
	Finish Board (Walls to 8' - Screwed On)			
	3/8" 4' x 8' and 12'	SqFt	.34	.22
	1/2" 4' x 8' and 12'	SqFt	.35	.24
	5/8" 4' x 8' and 12	SqFt	.36	.27
	1" 4' x 8' and 12' Core Board	SqFt	.37	.50
	1" 4' x 8' and 12' Bead Board	SqFt	.35	.50
	Add for Adhesive Application	SqFt	.08	.02
	Deduct for Nail-on Application	SqFt	.02	-
	Add for: Fire Resistant Board	SqFt	.02	.07
	Add for: Foil Backed (Insulating)	SqFt	.02	.10
	Add for: Moisture Resistant Type 5/8" & 1/2"	SqFt	.02	.06
	Add for: Work Above 8'	SqFt	.06	-
	Add for: Work Over 2-Story per Story	SqFt	.06	-
	Add for: Ceiling Work - To Wood	SqFt	.14	-
	To Steel	SqFt	.34	-
	Add for: Resilient Clip Application	SqFt	.33	.07
	Add for: Small Cut Up Areas	SqFt	.36	.05
	Add for: Beam and Column Work	SqFt	.46	.05
	Add for: Circular Construction	SqFt	.40	.07
	Add for: Vinyl Faced Board			
	3/8" x 4' x 8' - Cost Variable	SqFt	.40	.75
	1/2" x 4' x 8' - Cost Variable	SqFt	.44	.80
	Metal Framing (Studs, Runners and Channels)			
	Studs 25-Gauge 1 5/8"	LnFt	.50	.30
	2 1/2"	LnFt	.51	.32
	3 5/8"	LnFt	.58	.35
	6"	LnFt	.53	.47
	8"	LnFt	.58	.50
	Studs 209A 1 5/8"	LnFt	.48	.40
	2 1/2"	LnFt	.50	.45
	3 5/8"	LnFt	.52	.51
	6"	LnFt	.53	.68
	8"	LnFt	.57	.75
	7/8" Furring Channels	LnFt	.50	.32
	1 1/2" Furring Channels (Cold Rolled)	LnFt	.51	.45
	Resilient Channels	LnFt	.48	.35
	Add for Over 10' Lengths	LnFt	.17	.08
	Add for 20-Gauge 6" Material	LnFt	.16	.45
	Add for 16-Gauge 8" Material	LnFt	.28	1.00
	Metal Trim			
	Casing Bead	LnFt	.68	.22
	Corner Bead	LnFt	.52	.14
	L & J Bead 3/8" - 1/2" - 5/8"	LnFt	.52	.21
	Expansion Bead - Metal	LnFt	.52	.60
	Plastic	LnFt	.52	.30
	Metal Mouldings	LnFt	.80	.52
	Taping and Sanding	SqFt	.30	.07
	Texturing	SqFt	.38	.08
	Thin Coat Plaster – or Skin Coat	SqFt	.85	.28
	Sound Deadening Insulation - 2 1/2"	SqFt	.30	.25
	3 1/2"	SqFt	.35	.25
	6"	SqFt	.37	.32
.25	Cement Board 1/2" 3' x 5 1/2'	SqFt	.35	1.10
	1/2" 4' x 8'	SqFt	.35	1.10

0603.0 MILLWORK & CUSTOM WOODWORK (M) (Carpenters)

.1 CUSTOM CABINET WORK	UNIT	LABOR	MATERIAL Prefinished Red Oak or Birch	MATERIAL Plastic Laminate
See Division 11 - Stock Cabinets (Medium etc)				
Base Cabinets 35" High x 24" Deep -				
18" W	LnFt	17.40	109.00	104.00
24" W	LnFt	17.40	104.00	99.00
30" W	LnFt	16.40	100.00	95.00
36" W	LnFt	16.40	95.00	90.00
36" Wide Sink Fronts	LnFt	15.30	68.00	63.00
42"	LnFt	15.30	73.00	68.00
36" Wide Corner Cabinet	LnFt	16.30	120.00	115.00
Add for Lazy Susan	Each	21.80	110.00	105.00
Add per Drawer	Each	3.50	23.00	22.00
Upper Cabinets - 12" Deep				
30" High 18" Wide	LnFt	17.40	79.00	76.00
24" W	LnFt	17.40	74.00	71.00
30" W	LnFt	16.40	72.00	69.00
36" W	LnFt	16.40	69.00	66.00
24" High 18" Wide	LnFt	16.40	68.00	65.00
24" W	LnFt	16.40	63.00	60.00
30" W	LnFt	16.40	61.00	58.00
36" W	LnFt	15.30	60.00	57.00
15" High 18" Wide	LnFt	15.30	62.00	59.00
24" W	LnFt	15.30	59.00	56.00
30" W	LnFt	14.10	58.00	55.00
36" W	LnFt	14.10	57.00	54.00
Utility Cabinets - 84" H x 24" D				
18" Wide	LnFt	20.75	160.00	138.00
24" W	LnFt	20.75	150.00	135.00
Add - Prefinished Interior	-	-	8%	-
Deduct - Prefinished to Unfinished	-	-	12%	-
Add for Plastic Lam. Interior	-	-	-	100%
Add to Above for Special Hdwe.	Unit	7.60	-	22.00
Add - White/Birch/White Oak/Maple	-	-	-	20%
Add - Walnut	-	-	-	100%
China or Corner Cabinet 84"H x 36"W	Each	110.00	625.00	620.00
Oven Cabinet 84" High x 27" High	Each	105.00	490.00	480.00
Vanity Cabinets 30" High x 21" Deep				
30" Wide	LnFt	32.70	105.00	102.00
36" W	LnFt	33.75	102.00	99.00
48" W	LnFt	37.00	98.00	95.00

.2 COUNTER TOPS, 25" with Back Splash 4"	UNIT	LABOR	MATERIAL
Plastic Laminated 1 1/2"	LnFt	10.40	29.00
Deduct for No Back Splash	LnFt	2.30	5.00
Plastic Laminated Vanity Top	LnFt	9.30	32.00
Marble	LnFt	9.85	62.00
Artificial	LnFt	9.85	55.00
Wood Cutting Block, 1 1/2"	LnFt	9.90	64.00
Stainless Steel	LnFt	11.50	102.00
Polyester-Acrylic Solid Surface	LnFt	10.40	120.00
Granite - Artificial - 1 1/4"	LnFt	12.60	65.00
3/4"	LnFt	11.50	49.00
Quartz	LnFt	11.50	60.00
Plastic (Polymer)	LnFt	11.00	40.00

0603.0 MILLWORK & CUSTOM WOODWORK, Cont'd...

.3 CUSTOM DOOR FRAMES & TRIM

	Unit	Labor	Birch	Fir	Poplar	Oak	Pine	Maple
Custom Jamb & Stop								
4 5/8" x 2'6" x 3/4" x 6'8"	Each	$40	$118	$62	$61	$110	$110	-
x 2'8" x 3/4"	Each	42	120	64	62	112	110	-
x 3'0" x 3/4"	Each	43	116	69	63	113	115	-
5 1/4" x 2'6" x 3/4" x 6'8"	Each	42	128	73	63	117	112	-
x 2'8" x 3/4"	Each	43	130	75	65	120	113	-
x 3'0" x 3/4"	Each	44	134	78	68	135	120	-
6 3/4" x 2'6" x 3/4" x 6'8"	Each	44	175	-	-	176	133	-
x 2'8" x 3/4"	Each	45	180	-	-	178	134	-
x 3'0" x 3/4"	Each	46	182	-	-	180	137	-
Stock Jamb & Stop								
4 5/8" x 2'6" x 3/4" x 6'8"	Each	38	105	36	-	103	66	-
x 2'8" x 3/4"	Each	39	107	37	-	105	69	-
x 3'0" x 3/4"	Each	38	109	38	-	108	70	-
5 1/4" x 2'6" x 3/4" x 6'8"	Each	39	115	42	50	115	71	-
x 2'8" x 3/4"	Each	40	117	43	51	117	72	-
x 3'0" x 3/4"	Each	42	122	44	52	127	74	-
6 3/4" x 2'6" x 3/4" x 6'8"	Each	42	144	-	-	160	100	-
x 2'8" x 3/4"	Each	43	146	-	-	162	102	-
x 3'0" x 3/4"	Each	44	152	-	-	164	105	-
Fire Rated Jamb & Stop								
4 5/8" x 3'0" x 1 1/16"	Each	43	185	-	-	170	-	-
5 1/4" x 3'0" x 1 1/16"	Each	44	193	-	-	181	-	-
6 3/4" x 3'0" x 1 1/16"	Each	46	198	-	-	186	-	-
Trim Only -11/16" x 2 1/4"	Each	39	112	30	40	97	79	-
(2 sides) - 3/4" x 3 1/2"	Each	46	180	36	56	127	103	-

.4 MOULDINGS & TRIM-STOCK

	Unit	Labor	Birch	Fir	Poplar	Oak	Pine	Maple
Apron 7/16" x 2"	LnFt	.85	1.32	-	.47	1.00	.69	-
11/16" x 2 1/2"	LnFt	1.02	1.86	-	-	1.55	1.27	1.38
Astragal 1 3/4" x 2 1/4"	LnFt	1.07	-	-	-	5.40	4.43	-
Base 7/16" x 2 3/4"	LnFt	.85	1.86	.65	.76	1.32	1.07	1.39
9/16" x 3 1/4"	LnFt	.90	3.10	-	1.03	2.37	1.37	1.50
Base Shoe 7/16" x 2 3/4"	LnFt	.85	-	-	.45	.76	.48	.77
Batten Strip 5/8" x 1 5/8"	LnFt	.75	-	-	-	1.16	.95	-
Brick Mould 1 1/4" x 3"	LnFt	1.12	-	1.05	-	-	1.50	-
Casing 11/16" x 2 1/4"	LnFt	.90	1.44	.78	.70	1.10	1.10	1.06
3/4" x 3 1/2"	LnFt	.97	2.55	.88	.75	2.12	1.93	-
Chair Rail 5/8" x 1 3/4"	LnFt	.78	1.85	-	1.18	1.75	1.42	-
or Dado 11/16" x 2 1/4"	LnFt	.83	2.49	-	-	2.18	1.44	-
Closet Rod 1 5/16"	LnFt	1.25	-	.88	-	1.30	-	-
Corner Bead 3/4" x 3/4"	LnFt	.75	-	-	.64	1.13	.87	1.11
1 1/8" x 1 1/8"	LnFt	.82	-	-	1.12	1.65	1.50	1.56
Cove Mould 3/4" x 7/8"	LnFt	1.13	1.22	-	.54	.86	.69	.96
Crown & Cove 9/16" x 2 3/4"	LnFt	1.17	2.92	-	1.07	1.86	1.66	2.04
9/16" x 3 5/8"	LnFt	1.27	3.60	-	-	3.07	2.18	-
11/16" x 4 5/8"	LnFt	1.35	-	-	-	3.60	3.30	-
Drip Cap 11/16" x 1 3/4"	LnFt	.93	-	-	-	-	1.48	-
Half Round 1/2" x 1"	LnFt	1.06	-	-	-	-	1.06	-
Hand Rail 1 5/8" x 1 3/4"	LnFt	1.48	-	1.42	-	-	1.58	-
Hook Strip 5/8" x 2 1/2"	LnFt	.88	-	.88	.92	-	1.04	-
Picture Mould 3/4" x 1 1/2"	LnFt	.88	.92	-	-	1.05	1.04	-
Quarter Round 3/4" x 3/4"	LnFt	.75	-	.40	.54	.86	.55	.87
1/2" x 1/2"	LnFt	.70	-	.31	-	-	.36	-
Sill Casement 3/4" x 1 7/8"	LnFt	1.20	-	-	-	-	1.48	-
Stool 11/16" x 2 1/2"	LnFt	1.20	3.40	1.12	1.37	2.94	1.87	-

Above prices based on 200 LnFt. Deduct for Over 200 LnFt - 20%.
Add for Walnut Trim - 100% Add for Custom Trim - 50%

0603.0 MILLWORK & CUSTOM WOODWORK, Cont'd...

		Unit	Labor				Oak	Pine	Walnut
.5	STAIRS								
	Treads 1 1/6" x 10 1/4"	LnFt	1.85				4.45	3.50	-
	Risers 3/4" x 7 1/2"	LnFt	1.65				4.90	2.50	-
	Skirt Bds 9/16" x 9 1/2"	LnFt	2.10				6.30	2.60	-
	Nosings 1/8" x 3 1/2"	LnFt	1.55				3.15	1.85	-
.6	SHELVING								
	12" Deep	SqFt	4.65	31.70	-	-	24.21	12.20	72.00
	8"	SqFt	4.70	26.60	-	-	22.15	11.15	59.00
.7	CUSTOM PANELING	SqFt	2.85	18.70	-	-	13.39	-	22.00
.8	THRESHOLDS 3/4" x 3 1/2"								
	Standard	LnFt	6.90	-	-	-	5.50	-	-
	With Vinyl	LnFt	6.90	-	-	-	6.00	-	-

0604.0 GLUE LAMINATED (M) (Carpenters) (F.O.B. Cars)

		UNIT	LABOR	MATERIAL
.1	ARCHES AND RIGID FRAMES: Arches	MBF	610.00	1,800.00
	Rigid Frames	MBF	620.00	1,960.00
.2	BEAMS, JOISTS & PURLINS: 4" x 6", 8"& 10"	MBF	620.00	2,270.00
	6" x 6", 8" & 10"	MBF	600.00	2,140.00
	8" x 6", 8" & 10"	MBF	590.00	2,050.00
.3	DECKS: Doug.Fir - 3" x 6" (2 1/4"x 5 1/4")	SqFt	1.80	5.00
	4" x 6" (3" x 5 1/4")	SqFt	1.85	5.40
	Cedar - 3" x 6" (2 1/4"x 5 1/4")	SqFt	1.90	5.90
	4" x 6"	SqFt	2.00	6.30
	Pine - 3" x 6" (2 1/4"x 5 1/4")	SqFt	1.80	4.25

0605.0 PREFABRICATED WOOD & PLYWOOD COMPONENTS (See 0601.22, Job Fabricated)

.1	WOOD TRUSSED RAFTERS			
	4 -12 Pitch 24" O.C. With 2' Overhang 57 lb. Loading			
	16' Span	Each	31.60	63.00
	20'	Each	33.75	65.50
	22'	Each	34.20	67.40
	24'	Each	37.00	70.80
	26'	Each	39.00	77.50
	28'	Each	42.50	91.70
	30'	Each	46.60	103.00
	32'	Each	52.50	114.30
	Add for 5-12 Pitch	Each	5%	15%
	Add for 6-12 Pitch	Each	12%	25%
	Add for Gable End Type (Flying)	Each	19.60	34.00
	Add for 16" O.C.	Each	-	10%
	Add for Hip Ends - 24' Span	Each	152.00	390.00
	Add for Energy Type	Each	-	10%
.2	WOOD FLOOR AND ROOF TRUSS JOISTS			
.21	Open Wood Web - Gusset Plate Connection			
	24" O.C. Up to 23' Span x 12" - Single Chord	LnFt	1.45	4.80
	to 24' Span x 15"	LnFt	1.50	4.90
	to 27' Span x 18"	LnFt	1.55	5.00
	to 30' Span x 21"	LnFt	1.70	5.20
	to 25' Span x 15" - Double Chord	LnFt	1.60	4.45
	to 30' Span x 18"	LnFt	1.95	4.50
	to 35' Span x 21"	LnFt	2.25	4.70
	Open Metal Web - Pin Connection			
	to 25' Span x 18"	LnFt	1.53	4.50
	to 30' Span x 22"	LnFt	1.64	4.60
	to 35' Span x 30"	LnFt	1.85	4.80
	to 40' Span x 30"	LnFt	2.30	4.95

		UNIT	LABOR	MATERIAL
0605.0	**PREFABRICATED WOOD & PLYWOOD COMPONENTS, Cont'd...**			
.22	Plywood Web			
	24" O.C. Up to 15' Span x 9 1/2"	LnFt	1.22	2.80
	to 19' Span x 11 7/8"	LnFt	1.28	2.90
	to 21' Span x 14"	LnFt	1.43	3.25
	to 22' Span x 16"	LnFt	1.60	3.40
.3	LAMINATED VENEER STRUCTURAL BEAMS			
.31	Micro Lam (Plywood) 7 1/4" x 1 3/4"	LnFt	1.25	3.00
	9 1/2" x 1 3/4"	LnFt	1.30	4.30
	11 7/8" x 1 3/4"	LnFt	1.37	4.90
	14" x 1 3/4"	LnFt	1.45	5.90
	16" x 1 3/4"	LnFt	1.55	6.60
	18" x 1 3/4"	LnFt	1.73	7.30
.32	Glue Lam (Dimension Lumber)			
	9" x 3 1/2" Industrial	LnFt	1.30	9.10
	12"x 3 1/2"	LnFt	1.37	11.00
	15"x 3 1/2"	LnFt	1.45	13.00
	9" x 5 1/8"	LnFt	1.30	12.50
	12"x 5 1/8"	LnFt	1.48	15.50
	15"x 5 1/8"	LnFt	1.65	19.50
	18"x 5 1/8"	LnFt	1.80	24.00
	24"x 5 1/8"	LnFt	2.15	30.00
	Add for Architectural Grade	LnFt	.24	10%
	Add for 6 3/4"	LnFt	.25	5.00
	Add for Pine Treated	LnFt		4.00
.4	DECKS - 4' x 30' with Plywood 2 Sides - Insulated	SqFt	1.10	6.10
0606.0	**WOOD TREATMENTS (Preservatives)**			
	SALTS - PRESSURE (Water Borne)			
	Dimension 2 x 4-2 x 6 and 2 x 8, 2 x 10-2 x 12	BdFt	-	.16
	2 x 10 and 2 x 12	BdFt	-	.25
	Plywood 5/8" CDX	SqFt	-	.22
	SALTS - NON PRESSURE	BdFt	-	.12
	CREOSOTE (Penta)	BdFt	-	.33
	FIRE RETARDANTS - PRESSURE			
	Dimension and Timbers	BdFt	-	.32
	Plywood 5/8" CDX	SqFt	-	.32
0607.0	**ROUGH HARDWARE**			
.1	NAILS, BOLTS, ETC. Job Average	SqFt & BdFt	-	.02
	Common (50# Box) 8 Penny	CWT	-	50.00
	16 Penny	CWT	-	45.00
	Finish #3	CWT	-	59.00
	Add for Coated	CWT	-	60%
.2	JOIST HANGERS	Each	1.43	1.00
.3	BOLTS - 5/8" x 12"	Each	1.72	1.50
0608.0	**EQUIPMENT (Saws, Woodworking, Drills, Etc.)**			
	Average Job - As a Percentage of Labor	-	-	3%
	Heavy Equipment in Division 2			
	See 1-6A and 1-7A for Rental Rates			

Recommend 10% Decrease on Labor Units for Residential Construction

The costs below are average, priced as total contractor/subcontractor costs with an overhead of 5% and fee of 10% included. Units include fasteners and equipment, 35% taxes and insurance on labor, and 10% General Conditions and Equipment. Also included are nails, cutting, size and lap allowances. See Example on Page 4.

0601.0 ROUGH CARPENTRY

		UNIT	COST
.1	LIGHT FRAMING AND SHEATHING		
.11	Joists and Headers - Floor Area - 16" O.C.		
	2" x 6" Joists - with Headers & Bridging	SqFt	2.05
	2" x 8" Joists	SqFt	2.55
	2" x 10" Joists	SqFt	2.90
	2" x 12" Joists	SqFt	3.60
	Add for Ceiling Joists, 2nd Floor and Above	SqFt	8%
	Add for Sloped Installation	SqFt	15%
.12	Studs, Plates and Framing - 8' Wall Height - 16" O.C.		
	2" x 3" Stud Wall - Non-Bearing (Single Top Plate)	SqFt	1.32
	2" x 4" Stud Wall - Bearing (Double Top Plate)	SqFt	1.50
	2" x 4" Stud Wall - Non-Bearing (Single Top Plate)	SqFt	1.40
	2" x 6" Stud Wall - Bearing (Double Top Plate)	SqFt	1.85
	2" x 6" Stud Wall - Non-Bearing (Single Top Plate)	SqFt	1.80
	Add for Stud Wall - 12" O.C.	SqFt	8%
	Deduct for Stud Wall - 24" O.C.	SqFt	22%
	Add for Bolted Plates or Sills	SqFt	.10
	Add for Each Foot Above 8'	SqFt	.06
	Add to Above for Fire Stops, Fillers and Nailers	SqFt	.45
	Add for Soffits and Suspended Framing	SqFt	.60
.13	Bridging - 1" x 3" Wood Diagonal	Set	2.40
	2" x 8" Solid	Each	2.60
.14	Rafters		
	2" x 4" Rafter (Incl Bracing) 3 - 12 Slope	SqFt	1.75
	4 - 12 Slope	SqFt	1.90
	5 - 12 Slope	SqFt	2.00
	6 - 12 Slope	SqFt	2.30
	Add for Hip-and-Valley Type	SqFt	.42
	Add for 1' 0" Overhang - Total Area of Roof	SqFt	.15
	2" x 6" Rafter (Incl Bracing) 3-12 Slope	SqFt	2.20
	4-12 Slope	SqFt	2.25
	5-12 Slope	SqFt	2.30
	6-12 Slope	SqFt	2.40
	Add for Hip - and - Valley Type	SqFt	.52
.15	Stairs	Each Floor	435.00
.16	Sub Floor Sheathing (Structural)		
	1" x 8" and 1" x 10" #3 Pine	SqFt	1.57
	1/2" x 4' x 8' CD Plywood - Exterior	SqFt	1.30
	5/8" x 4' x 8'	SqFt	1.46
	3/4" x 4' x 8'	SqFt	1.67
.17	Floor Sheathing (Over Sub Floor)		
	3/8" x 4' x 8' CD Plywood	SqFt	1.22
	1/2" x 4' x 8'	SqFt	1.30
	5/8" x 4' x 8'	SqFt	1.42
	1/2" x 4' x 8' Particle Board	SqFt	1.05
	5/8" x 4' x 8'	SqFt	1.14
	3/4" x 4' x 8'	SqFt	1.34

0601.0	ROUGH CARPENTRY, Cont'd...	UNIT	COST
.18	Wall Sheathing		
	1' x 8" and 1" x 10" - #3 Pine	SqFt	1.58
	3/8" x 4' x 8' CD Plywood - Exterior	SqFt	1.20
	1/2" x 4' x 8'	SqFt	1.30
	5/8" x 4' x 8'	SqFt	1.44
	3/4" x 4' x 8'	SqFt	1.68
	25/32" x 4' x 8' Fiber Board - Impregnated	SqFt	.92
	1" x 2' x 8' T&G Styrofoam	SqFt	1.17
	2" x 2' x 8'	SqFt	1.72
.19	Roof Sheathing - Flat Construction		
	1" x 6" and 1" x 8" - #3 Pine	SqFt	1.55
	1/2" x 4' x 8' CD Plywood - Exterior	SqFt	1.33
	5/8" x 4' x 8'	SqFt	1.45
	3/4" x 4' x 8'	SqFt	1.70
	Add for Sloped Roof Construction (to 5-12 slope)	SqFt	.16
	Add for Steep Sloped Construction (over 5-12 slope)	SqFt	.28
	Add to Above Sheathing (0601.16 thru 0601.19)		
	AC or AD Plywood	SqFt	.28
	10' Length Plywood	SqFt	.21
.2	HEAVY FRAMING		
.21	Columns and Beams - 16' Span Average - Floor Area	SqFt	6.80
	20' Span	SqFt	7.05
	24' Span	SqFt	8.20
.22	Deck 2" x 6" T&G - Fir Random Construction Grade	SqFt	4.00
	3" x 6" T&G - Fir Random Construction Grade	SqFt	5.25
	4" x 6"	SqFt	6.80
	2" x 6" T&G - Red Cedar	SqFt	5.05
	Add for D Grade Cedar	SqFt	1.70
	2" x 6" T&G - Panelized Fir	SqFt	3.70
.3	MISCELLANEOUS CARPENTRY		
.31	Blocking and Bucks (2" x 4" and 2" x 6")	EACH	or BdFt
	Doors & Windows - Nailed to Concrete or Masonry	47.15	2.40
	Bolted to Concrete or Masonry	53.30	3.30
	Roof Edges - Nailed to Wood	-	1.85
	Bolted to Concrete (Incl. bolts)	-	2.80
.32	Grounds and Furring	LnFt	or BdFt
	1" x 6" Fastened to Wood	1.05	2.10
	1" x 4"	.90	2.70
	1" x 3"	.85	3.40
	1" x 3" Fastened to Concrete/Masonry - Nailed	1.30	5.20
	Gun Driven	1.22	4.88
	PreClipped	1.50	6.00
	2" x 2" Suspended Framing	1.35	4.05
	Not Suspended Framing	1.13	3.39
.33	Cant Strips		
	4" x 4" Treated and Nailed	1.55	-
	Treated and Bolted (Including Bolts)	2.25	-
	6" x 6" Treated and Nailed	3.10	-
	Treated and Bolted (Including Bolts)	4.00	-
.34	Building Papers and Sealers	UNIT	COST
	15" Felt	SqFt	.15
	Polyethylene - 4 mil	SqFt	.14
	6 mil	SqFt	.16
	Sill Sealer	SqFt	.70

0602.0 FINISH CARPENTRY

		TO	%	
.1	FINISH SIDINGS AND FACING MATERIALS (Exterior)	WEATHER	ADDED	COST
.11	Boards and Beveled Sidings			
	Cedar Beveled - Clear Heart 1/2" x 4"	2 3/4"	51%	5.25
	1/2" x 6"	4 3/4"	31%	4.60
	1/2" x 8"	6 3/4"	23%	3.90
	Rough Sawn 7/8" x 8"	6 3/4"	23%	4.00
	7/8" x 10"	8 3/4"	19%	4.12
	7/8" x 12"	10 3/4"	17%	4.13
	3/4" x 12"	-	-	3.55
	Redwood Beveled - Clear Heart			
	1/2" x 6"	4 3/4"	31%	5.00
	1/2" x 8"	6 3/4"	23%	4.95
	5/8" x 10"	8 3/4"	19%	5.10
	3/4" x 6"	4 3/4"	31%	5.20
	3/4" x 8"	6 3/4"	23%	5.80
	3/4" x 6" Board - Clear -		Varies	3.00
	3/4" Tongue & Groove -		Varies	4.75
	1" Rustic Beveled - 6" and 8"		Varies	5.25
	5/4" Rustic Beveled - 6" and 8"		Varies	5.15

			UNIT	COST
	Add for Metal Corners		Each	1.40
	Add for Mitering Corners		Each	2.90
.12	Plywood Fir AC Smooth -One Side 1/4"		SqFt	1.80
	3/8"		SqFt	1.93
	1/2"		SqFt	2.00
	5/8"		SqFt	2.05
	3/4"		SqFt	2.30
	5/8" - Grooved and Rough Faced		SqFt	2.35
	Cedar - Rough Sawn 3/8"		SqFt	2.25
	5/8"		SqFt	2.40
	3/4"		SqFt	2.80
	5/8" - Grooved and Rough Faced		SqFt	2.85
	Add for Wood Batten Strips, 1" x 2" - 4' O.C.		SqFt	.35
	Add for Splines		SqFt	.38
.13	Hardboard - Paneling and Lap Siding (Primed)			
	3/8" Rough Textured Paneling -	4' x 8'	SqFt	1.85
	7/16" Grooved -	4' x 8'	SqFt	1.80
	7/16" Stucco Board Paneling		SqFt	1.98
	7/16" x 8" Lap Siding		SqFt	2.30
	7/16" x 12" Lap Siding		SqFt	2.18
	Add for Pre-Finishing		SqFt	:17
.14	Shingles			
	Wood- 16" Red Cedar - 12" to Weather	- #1	SqFt	3.65
		- #2	SqFt	2.90
		- #3	SqFt	2.75
	24" - 1/2" to 3/4" Red Cedar Hand Splits	- #1	SqFt	3.140
	24" - 3/4" to 1 1/4" Red Cedar Hand Splits	- #1	SqFt	3.50
		- #2	SqFt	3.14
		- #3	SqFt	2.93
	Add for Fire Retardant		SqFt	.43
	Add for 3/8" Backer Board		SqFt	.75
	Add for Metal Corners		Each	1.38
	Add for Ridges, Hips and Corners		LnFt	3.82
	Add for 8" to Weather		SqFt	25%

		COST LN FT	or	COST SQ FT
0602.0	**FINISH CARPENTRY, Cont'd...**			
.15	Facia - Pine #2 1" x 8"	1.75		2.62
	Cedar #3 1" x 8"	2.60		3.90
	Redwood - Clear 1" x 8"	3.80		5.70
	Plywood 5/8" - ACX 1" x 8"	1.80		2.70
.16	Soffit - Plywood 1/2" - ACX 1/2" Fir ACX			2.60
	Aluminum			2.95
	Plastic (Egg Crate)			3.15
.2	FINISH WALLS (Interior) (15% Added for Size & Cutting)			
.21	Boards, Cedar - #3 1" x 6" and 1" x 8"			3.83
	Knotty 1" x 6" and 1" x 8"			4.00
	D Grade 1" x 6" and 1" x 8"			4.78
	Aromatic 1" x 6" and 1" x 8"			4.95
	Redwood - Construction 1" x 6" and 1" x 8"			4.15
	Clear 1" x 6" and 1" x 8"			5.50
	Fir - Beaded 5/8" x 4"			3.70
	Pine - #2 1" x 6" and 1" x 8"			2.90
.22	Hardboard (Paneling) Tempered - 1/8"			1.18
	1/4"			1.34
.23	Plywood (Prefinished Paneling) 1/4" Birch - Natural			2.83
	3/4" Birch - Natural			3.72
	1/4" Birch - White			3.40
	1/4" Oak - Rotary Cut			2.73
	3/4"			3.85
	1/4" Oak - White			3.69
	1/4" Mahogany (Lauan)			1.95
	3/4"			2.89
	3/4" Mahogany (African)			4.90
	1/4" Walnut			4.40
.24	Gypsum Board (Paneling) - See 0902			

		UNIT	COST Prefinished Red Oak or Birch	COST Plastic Laminate
0603.0	**MILLWORK & CUSTOM WOODWORK**			
.1	CUSTOM CABINET WORK (Red Oak or Birch)			
	Base Cabinets - Avg. 35" H x 24" D with Drawer	LnFt	177.00	174.00
	Sink Fronts	LnFt	106.00	105.00
	Corner Cabinets	LnFt	232.00	228.00
	Add per Drawer	LnFt	45.00	45.00
	Upper Cabinets - Avg. 30" High x 12" Deep	LnFt	133.00	132.00
	Utility Cabinets - Avg. 84" High x 24" Deep	LnFt	345.00	333.00
	China or Corner Cabinets - 84" High	Each	645.00	634.00
	Oven Cabinets - 84" High x 24" Deep	Each	435.00	421.00
	Vanity Cabinets - Avg. 30" High x 21" Deep	Each	230.00	224.00
	Deduct for Prefinishing Wood	Each	12%	-
.2	COUNTER TOPS - 25"			
	Plastic Laminated with 4" Back Splash	LnFt	-	52.00
	Deduct for No Back Splash	LnFt	-	9.50
	Granite - 1 1/4" - Artificial	LnFt	-	101.00
	3/4" - Artificial	LnFt	-	82.00
	Marble	LnFt	-	94.00
	Artificial	LnFt	-	86.00
	Wood Cutting Block	LnFt	-	97.00
	Stainless Steel	LnFt	-	153.00
	Polyester-Acrylic Solid Surface	LnFt	-	167.00
	Quartz	LnFt	-	90.00
	Plastic (Polymer)	LnFt	-	65.00
	(See 1103 for Stock Cabinets)			

0603.0 MILLWORK & CUSTOM WOODWORK, Cont'd...

.3	CUSTOM DOOR FRAMES - Including 2 Sides Trim	UNIT	COST
	Birch	Each	280.00
	Fir	Each	173.00
	Poplar	Each	178.00
	Oak	Each	220.00
	Pine	Each	210.00
	Walnut	Each	405.00

See 0802.87 for Stock Prehung Units

		LINEAR FOOT				
.4	MOULDINGS AND TRIM	Birch	Fir	Poplar	Oak	Pine
	Apron 7/16" x 2"	3.02	-	1.97	2.58	2.06
	Astragal 1 3/4" x 2 1/4"	-	-	-	6.50	4.98
	Base 7/16" x 2 3/4"	3.47	1.80	2.02	2.66	3.02
	9/16" x 3 1/4"	4.60	-	-	4.05	4.13
	Base Shoe 7/16" x 2 3/4"	-	-	1.76	2.38	2.48
	Batten Strip 5/8" x 1 5/8"	-	1.43	-	2.65	2.16
	Brick Mould 1 1/14" x 2"	-	2.50	-	-	2.50
	Casing 11/16" x 2 1/4"	3.20	1.95	2.15	3.00	2.73
	3/4" x 3 1/2"	4.37	2.50	2.20	4.30	3.78
	Chair Rail 5/8" x 1 3/4"	3.48	-	2.70	3.49	2.75
	Closet Rod 1 5/16"	-	2.55	-	-	3.15
	Corner Bead 1 1/8" x 1 1/8"	-	2.22	2.75	3.40	3.90
	Cove Moulding 3/4" x 3/4"	2.90	-	2.35	2.80	2.35
	Crown Moulding 9/16" x 3 5/8"	5.28	-	3.43	4.30	3.62
	11/16" x 4 5/8"	6.12	-	-	6.80	5.75
	Drip Cap	-	-	-	-	2.85
	Half Round 1/2" x 1"	-	-	-	-	3.32
	Hand Rail 1 5/8" x 1 3/4"	-	3.82	4.18	6.70	3.70
	Hook Strip 5/8" x 2 1/2"	-	2.07	2.37	-	2.48
	Picture Mould 3/4" x 1 1/2"	2.10	-	-	3.04	2.75
	Quarter Round 3/4" x 3/4"	-	1.23	1.55	2.44	1.92
	1/2" x 1/2"	-	1.33	-	-	1.52
	Sill 3/4" x 2 1/2"	-	2.50	-	-	3.32
	Stool 11/16 x 2 1/2"	5.33	2.50	3.70	5.85	3.98

		UNIT	Birch	Oak	Pine
.5	STAIRS (Treads, Risers, Skirt Boards)	LnFt	37.00	29.00	22.00
.6	SHELVING (12" Deep)	SqFt	40.00	35.00	24.00
.7	CUSTOM PANELING	SqFt	26.50	20.00	16.50
.8	THRESHOLDS - 3/4" x 3 1/2"	LnFt	-	19.00	-

0604.0 GLUE LAMINATE

	UNIT	COST
Arches 80' Span	SqFt	9.30
Beams & Purlins 40' Span	SqFt	6.40
Deck Fir - 3" x 6"	SqFt	8.05
4" x 6"	SqFt	8.90
Deck Cedar - 3" x 6"	SqFt	9.40
4" x 6"	SqFt	10.10
Deck Pine - 3" x 6"	SqFt	7.35

0605.0 PREFABRICATED WOOD & PLYWOOD COMPONENTS

			UNIT	COST		UNIT	COST
.1	WOOD TRUSSED RAFTERS - 24" O.C. - 4 -12 Pitch						
	Span To 16'	w/ Supports	Each	130.00	or	SqFt	4.05
	20'	and	Each	138.00		SqFt	3.45
	22'	2' Overhang	Each	141.00		SqFt	3.20
	24'		Each	146.00		SqFt	3.04
	26'		Each	152.00		SqFt	2.92
	28'		Each	168.00		SqFt	3.00
	30'		Each	198.00		SqFt	3.30
	32'		Each	220.00		SqFt	3.44
	Add for 5-12 Pitch		Each	5%		SqFt	15%
	Add for 6-12 Pitch		Each	12%		SqFt	25%
	Add for Scissor Truss		Each	28.00		SqFt	.22
	Add for Gable End 24'		Each	60.00		SqFt	.40
.2	WOOD FLOOR TRUSS JOISTS (TJI)						
	Open Web - Wood or Metal - 24" O.C.						
	Up To 23' Span x 12"	Single	LnFt	8.00	or	SqFt	4.00
	To 24' Span x 15"	Cord	LnFt	8.20		SqFt	4.10
	To 27' Span x 18"		LnFt	8.42		SqFt	4.21
	To 30' Span x 21"		LnFt	8.86		SqFt	4.43
	To 25' Span x 15"	Double	LnFt	7.88		SqFt	3.98
	To 30' Span x 18"	Cord	LnFt	8.20		SqFt	4.10
	To 36' Span x 21"		LnFt	9.15		SqFt	4.58
	Plywood Web - 24" O.C.						
	Up To 15' x 9 1/2"		LnFt	5.09		SqFt	2.54
	To 19' x 11 7/8"		LnFt	5.15		SqFt	2.58
	To 21' x 14"		LnFt	5.77		SqFt	2.88
	To 22' x 16"		LnFt	6.60		SqFt	3.20
.3	LAMINATED VENEER STRUCTURAL BEAMS						
.31	Micro Lam - Plywood - 24" O.C.						
	9 1/2" x 1 3/4"		LnFt	7.10	or	SqFt	3.55
	11 7/8" x 1 3/4"		LnFt	7.80		SqFt	3.90
	14" x 1 3/4"		LnFt	9.15		SqFt	4.58
	16" x 1 3/4"		LnFt	10.20		SqFt	5.10
.32	Glue Lam- Dimension Lumber- 24" O.C.						
	9" x 3 1/2"		LnFt	13.40		SqFt	6.70
	12" x 3 1/2"		LnFt	15.40		SqFt	7.70
	15" x 3 1/2"		LnFt	17.20		SqFt	8.60
	9" x 5 1/2"		LnFt	17.40		SqFt	8.70
	12" x 5 1/2"		LnFt	21.40		SqFt	10.70
	15" x 5 1/2"		LnFt	26.90		SqFt	13.45
	18" x 5 1/2"		LnFt	32.60		SqFt	16.30
	Add for Architectural Grade		LnFt	25%			
	Add for 6 3/4"		LnFt	40%			

0606.0 WOOD TREATMENTS

		UNIT	COST
	PRESERVATIVES – PRESSURE TREATED		
	Dimensions	BdFt	.17
	Timbers	BdFt	.27
	FIRE RETARDENTS		
	Dimensions & Timbers	BdFt	.35

0607.0 ROUGH HARDWARE

		UNIT	COST
.1	NAILS	SqFt or BdFt	.02
.2	JOIST HANGERS	Each	3.00
.3	BOLTS - 5/8" x 12"	Each	3.60

		UNIT	LABOR	MATERIAL
0701.0	**WATERPROOFING (L&M) (Roofers)**			
.1	MEMBRANE			
	1 Ply Membrane - Felt 15#	SqFt	.52	.35
	Fabric and Felt	SqFt	.72	.40
	2 Ply Membrane - Felt 15#	SqFt	.88	.57
	Fabric and Felt	SqFt	1.03	.60
.2	HYDROLITHIC	SqFt	.90	.85
.3	ELASTOMERIC- Rubberized Asphalt .060M & Poly	SqFt	.90	.58
	Liquid	SqFt	1.01	.70
.4	METALLIC OXIDE- 3 Coat	SqFt	.83	1.80
	3 Coat with Sand & Cement Cover	SqFt	1.07	2.00
.5	VINYL PLASTIC	SqFt	.52	1.15
.6	BENTONITE - 3/8"	SqFt	.73	.65
0702.0	**DAMPPROOFING (Roofers or Laborers)**			
.1	BITUMINOUS (Asphalt)			
.11	Trowel Mastics			
	Asphalt 1/16" (12 Sq.Ft. Gal.)	SqFt	.32	.23
	1/8" (20 Sq.Ft. Gal.)	SqFt	.37	.28
.12	Spray or Brush Liquid			
	Asphalt - 1 Coat - Spray	SqFt	.22	.15
	Spray	SqFt	.28	.20
	Brush	SqFt	.33	.22
.13	Hot Mop			
	Asphalt - 1 Coat - Including Primer	SqFt	.52	.38
	Fibrous Asphalt - 1 Coat	SqFt	.58	.41
.14	Trowel Mastics & Pre-formed Vapor Barriers			
	Mastic Fabric Mastic	SqFt	.54	.78
	and Polyvinyl Chloride	SqFt	.52	.65
.2	CEMENTITIOUS - 1 Coat - 1/2"	SqFt	.32	.16
	2 Coat - 1"	SqFt	.43	.20
.3	SILICONE (Masonry) - 1 Coat	SqFt	.24	.17
	2 Coat	SqFt	.31	.28
	Add for Scaffold to Above Sections	SqFt	.23	.12
0703.0	**BUILDING INSULATION (S) (Carpenters)**			
.1	FLEXIBLE (Batts and Rolls) - Friction			
	Fiberglass			
	2 1/4" R 7.4	SqFt	.25	.18
	3 1/2" R 11.0	SqFt	.26	.20
	3 5/8" R 13	SqFt.	.27	.27
	6" R 19.0	SqFt	.29	.32
	9" R 30.0	SqFt	.34	.54
	Add for Paper Faced Batts	SqFt	.02	.03
	Add for Polyethylene Barrier 2 Mil	SqFt	.04	.04
	Add for Staggered Stud Installation	SqFt	.09	.03

0703.0 BUILDING INSULATION, Cont'd...

		UNIT	LABOR	MATERIAL
.2	RIGID			
	Fiberglass			
	1" R 4.35 3# Density	SqFt	.30	.33
	1 1/2" R 6.52 3# Density	SqFt	.32	.50
	2" R 8.70 3# Density	SqFt	.34	.65
	Add for 4.2# Density per Inch	SqFt	-	.07
	Add for 6.0# Density per Inch	SqFt	-	.27
	Add for Kraft Paper Backed	SqFt	-	.08
	Add for Aluminum Foil Backed	SqFt	-	.38
	Add for Fireproof Type	SqFt	-	10%
	Expanded Styrene			
	Molded, White (Bead Board)			
	1" x 16" x 8' R 5.40	SqFt	.30	.17
	1" x 24" x 8', T&G R 5.40	SqFt	.33	.24
	1 1/2" R 6.52	SqFt	.34	.26
	2" R 7.69	SqFt	.38	.34
	2" x 24" x 8', T&G R 7.69	SqFt	.38	.48
	Extruded, Blue			
	1" x 24" x 8' R 5.40	SqFt	.30	.39
	T&G	SqFt	.33	.49
	1 1/2" R 6.52	SqFt	.34	.49
	2" R 7.69	SqFt	.38	.51
	2" x 24" x 8', T&G R 7.69	SqFt	.38	.60
	Perlite			
	1" R 2.78	SqFt	.30	.37
	2" R 5.56	SqFt	.38	.75
	Urethane			
	1" R 6.67	SqFt	.32	.45
	2" R 13.34	SqFt	.38	.65
.3	LOOSE			
	Expanded Styrene 1" R 3.80	SqFt	.24	.22
	Fiberglass 1" R 2.20	SqFt	.25	.23
	Rockwool 1" R 2.90	SqFt	.26	.23
	Cellulose 1" R 3.70	SqFt	.24	.24
.4	FOAMED			
	Urethane per Inch or Bd.Ft.	SqFt	.45	.80
.5	SPRAYED			
	Fibered Cellulose per Inch or Bd.Ft.	SqFt	.42	.45
	Polystyrene per Inch or Bd.Ft.	SqFt	.50	.53
	Urethane per Inch or Bd.Ft.	SqFt	.55	.95
.6	ALUMINUM PAPER	SqFt	.18	.16
.7	VINYL FACED FIBERGLASS			
	2 5/8" x 63 1/2" x 96" R 10.0	SqFt	.40	.85
	Add to All Above for Ceiling Work	SqFt	.06	-
	Add to All Above for Clip On	SqFt	.07	.02
	Add to All Above for Scaffold	SqFt	.20	.12

See Division 3 for Cementitious Poured and Rigid Insulation
See Division 4 for Core and Cavity Filled Masonry
See Division 7 for Roof Insulation
See Division 15 for Mechanical Insulations

		UNIT	COST
0704.0	**ROOF AND DECK INSULATION (L&M) (Roofers)**		
.1	RIGID - NON RATED		
	Fiber Board 1/2" R 1.39	Sq	51.10
	3/4" R 2.09	Sq	60.20
	1" R 2.78	Sq	69.95
	Fiberglass 1" R 3.70	Sq	80.65
	1 1/2" R 2.09	Sq	106.00
	Perlite 1/2" R 1.39	Sq	50.20
	3/4" R 2.09	Sq	62.20
	1" R 2.78	Sq	73.10
	2" R 5.26	Sq	109.00
	Expanded Polystrene 1" R 4.35	Sq	67.05
	1 1/2" R 6.52	Sq	78.00
	2" R 7.69	Sq	97.60
.2	RIGID - FIRE RATED		
	Isocyanurate 1" R 6.30 Glass Faced	Sq	111.00
	1 1/5" R 7.60	Sq	116.30
	1 1/2" R 10.00	Sq	122.10
	1 3/4" R 12.00	Sq	133.10
	2" R 14.00	Sq	137.80
	Phenolic Foam 1" R 7.60	Sq	92.10
	1 1/2" R 12.50	Sq	103.40
	2" R 16.60	Sq	127.00
	Add for 5/8" Sheet Rock over Steel Deck	Sq	14.70
	Add for Fiber or Mineral Cants	LnFt	1.60
	Add for Felt Faces	Sq	10.45
	Add for Composite Insulations	Sq	25%
.3	SPRAYED - 2" Polystyrene	Sq	250.00
0705.0	**MEMBRANE ROOFING (L&M) (Roofers)**		
.1	BUILT-UP ROOFING (50 SQUARES OR MORE)		
	Asphalt & Gravel - 15# Glass/Organic Felts 3-Ply	Sq	150.30
	4 - Ply	Sq	162.00
	5 - Ply	Sq	182.90
	Add for Pitch and Gravel	Sq	44.20
	Add for Curbed Openings	Each	131.20
	Add for Round Vent Openings	Each	118.80
	Add for Roof Skids	LnFt	28.60
	Add for Sloped Roofs	Sq	33.75
	Add for Applications on Wood	Sq	11.50
	Add for Marble Chips	Sq	50.00
	Add for Barriers - 15" Felt	Sq	9.90
	Fire Resistant	Sq	22.60
	Add for Walkways - 1"	SqFt	4.80
	Add for Bond	Sq	11.00
	Add for Roof Openings Cut in Existing	Each	417.00

See Quick Estimating 7A-10 for Combined Roofing, Insulation and Sheet Metal

0706.0 SINGLE-PLY ROOFING No Insulation or Flashing Included

	UNIT	Gravel Ballast	Application Mechanical Fastened	Adhesive Fastened
Butylene 100 mil	Sq	185.00	200.00	205.00
EPDM 50 mil	Sq	145.00	175.00	200.00
Neoprene 60 mil	Sq	270.00	325.00	225.00
P.V.C. 50 mil	Sq	135.00	195.00	215.00
Add per Opening	Opng	130.00	-	-

0707.0 TRAFFIC ROOF COATINGS (L&M)

		UNIT	LABOR	MATERIAL
.1	PEDESTRIAN TRAFFIC TYPE	SqFt	-	6.50
.2	VEHICULAR TRAFFIC TYPE			
	Rubberized Coating with 3/4" Asphalt Topping	SqFt	-	4.40
	Polyurethane with Non-Slip Aggregates on 20 mil silicone rubber	SqFt	-	4.00
	Elastomeric with Non-Skid - 2 Coat	SqFt	-	2.60
	3 Coat	SqFt	-	3.35

0708.0 SHINGLE, TILE and CORRUGATED ROOFING

		UNIT	LABOR	MATERIAL
.1	ASPHALT SHINGLES (S) (Carpenters) 4-12 Pitch			
	Square Butt 235#- Seal Down	SqFt	.44	.40
	300# - Laminated	SqFt	.49	.60
	325# - Fire Resistant	SqFt	.52	.62
	Timberline 300# - 25 Year	SqFt	.59	.52
	300# - 30 Year	SqFt	.61	.65
	Tab Lock 340#	SqFt	.59	.90
	Roll 90# - 100 SqFt per Roll	SqFt	.27	.19
	55# - 100 SqFt per Roll	SqFt	.23	.16
	Base Starter - 240 SqFt per Roll	SqFt	.23	.23
	Ice & Water Starter - 100 SqFt per Roll	SqFt	.28	.42
	Add for Boston Ridge Shingle	LnFt	1.10	.48
	Add for 15# Felt Underlayment	SqFt	.06	.04
	Add for Pitches 6-12 & Up - Ea Pitch Increase	SqFt	.06	-
	Add for Removal and Haul - Away (2 Layers)	SqFt	.30	.14
	Add for Chimneys, Skylights & Bay Windows	Each	44.00	-
.2	FIBERGLASS SHINGLES - 215# - Seal Down	SqFt	.44	.40
	250# - Seal Down	SqFt	.47	.57
.3	WOOD			
	Sawed			
	Red Cedar - 16" x 5" to Weather - #1 Grade	SqFt	.79	1.60
	#2 Grade	SqFt	.77	1.25
	#3 Grade	SqFt	.72	.75
	Hand Splits			
	Red Cedar - 1/2" x 3/4", 24" x 10" to Weather	SqFt	.93	1.30
	3/4"x1 1/4", 24" x 10" to Weather	SqFt	1.01	1.60
	Fancy Butt - 7" to Weather	SqFt	1.10	3.40
	Add for Fire Retardant	SqFt	-	.42
	Add for Pitches over 5-12, Each Pitch Increase	SqFt	.07	-
.4	METAL			
	Aluminum - Mill .020	SqFt	.83	1.60
	.030	SqFt	.84	1.95
	Anodized .020	SqFt	.83	2.03
	.030	SqFt	.84	2.30
	Steel - Enameled - Colored	SqFt	.88	2.80
	Galvanized - Colored	SqFt	.88	2.13

0708.0 SHINGLE, TILE AND CORRUGATED ROOFING, Cont'd...

		UNIT	COST
.6	SLATE SHINGLES (Incl. Underlayment) (L&M) (Roofers)		
	Vermont and Pennsylvania		
	3/16" - 6 1/2" to 8 1/2" to Weather		
	Weathered Green or Black	Sq	1,050.00
	Gray	Sq	1,050.00
	Fading Purple	Sq	1,050.00
	Unfading Purple	Sq	1,370.00
	Add for Variegated Colors & Sizes - 1/4" to 1/2"	Sq	355.00
	3/8" to 1"	Sq	680.00
	Add for Hips and Valleys	LnFt	53.00
	Add for Pitches over 6-12	Sq	72.00
.7	TILE (L&M) (Roofers)		
.71	Clay, Interlocking Shingle Type - Designed	Sq	1,035.00
	Early American	Sq	1,040.00
	Williamsburg	Sq	1,150.00
	Mormon	Sq	1,660.00
	Architectural Pattern Shingle Type - Provincial	Sq	990.00
	Georgian	Sq	990.00
	Colonial	Sq	990.00
	Spanish, Red	Sq	710.00
	Mission	Sq	1,150.00
.72	Concrete - Plain - 10" to Weather - 9" Wide	Sq	630.00
	Colored	Sq	700.00
	Add for Staggered Installation	Sq	230.00
.73	Steel - 16 Ga - Ceramic Coated	Sq	430.00
.74	Aluminum .019	Sq	500.00
.8	CORRUGATED PANELS		
.81	Concrete - Plain	Sq	3.60
.82	Vinyl - 120	Sq	5.70
.83	Fiberglass - 8 oz	Sq	3.80
.84	Aluminum .024 - Plain	Sq	3.15
	.024 - Painted	Sq	3.85
	See Division 5 for Corrugated Metal Panels		

0709.0 SHEET METAL WORK (L&M)

	(Sheet Metal Workers)	UNIT	COST Galvanized 25 Ga	Aluminum .032"	Copper 16 oz	Enameled
.1	DOWNSPOUTS, GUTTERS AND FITTINGS					
	Downspouts - 3" x 2"	LnFt	2.50	2.50	6.84	3.12
	4" x 3"	LnFt	3.11	3.36	9.71	4.18
	5" x 4"	LnFt	4.28	4.35	10.65	5.35
	Round 3"	LnFt	2.40	2.95	6.30	3.11
	4"	LnFt	2.83	3.52	7.45	3.95
	5"	LnFt	3.57	4.27	8.32	4.74
	Chute Type 4" x 6"	LnFt	4.13	4.80	-	4.76
	5" x 7"	LnFt	4.73	5.40	-	6.57
	Gutters - Box or Half Round 4"	LnFt	3.30	4.28	6.83	5.01
	5"	LnFt	3.86	4.38	8.43	4.98
	6"	LnFt	4.73	5.03	8.63	5.27
	Add for Guards	LnFt	-	1.06	-	-
	Fittings (Elbows, Corners, etc.)	LnFt	9.95	9.40	11.70	11.70

See Quick Estimating 7A-10 for Combined Roofing, Insulation and Sheet Metal.

0709.0 SHEET METAL WORK, Cont'd...

	UNIT	COST Galva-nized 25 Ga	Alum-inum .032"	Copper 16 oz	Enameled	PVC
.2 FASCIAS, GRAVEL STOPS & TRIM						
Fascias - 6"	LnFt	4.70	7.25	10.40	5.90	6.32
8"	LnFt	5.04	8.12	11.60	6.21	6.90
Copings - 12"	LnFt	5.95	9.57	12.05	7.25	-
Scuppers	Each	112.00	112.00	134.00	117.00	-
Gravel Stops - 4"	LnFt	4.60	6.75	7.92	5.85	-
6"	LnFt	4.28	7.58	9.85	6.03	6.52
8"	LnFt	5.55	8.00	11.15	6.78	6.95
Add for Durodonic Finish	LnFt		25%			

	UNIT	COST
.3 FLASHINGS		
Galvanized - 25 ga	SqFt	4.72
Aluminum - .019	SqFt	4.40
.032	SqFt	4.72
Copper - 16 oz	SqFt	7.65
20 oz	SqFt	7.85
Asphalted Fabric - Plain	SqFt	2.35
Copper Backed	SqFt	3.84
Aluminum Back	SqFt	3.38
Vinyl Fabric - .020	SqFt	1.25
.030	SqFt	1.42
Lead - 2.5#	SqFt	7.13
Rubber - 1/16"	SqFt	7.83
PVC - 24 mil	SqFt	3.20
.4 SHEET METAL ROOFING - Including Underlayment		
Copper - Standing Seam - 16 oz	Sq	1,095.00
20 oz	Sq	1,120.00
Batten Seam 16 oz	Sq	995.00
20 oz	Sq	1,155.00
Flat Lock 16 oz	Sq	920.00
20 oz	Sq	1,060.00
Stainless Steel - Batten Seam 28 ga	Sq	1,010.00
Standing Seam 28 ga	Sq	1,120.00
Flat Seam 28 ga	Sq	953.00
Monel - Batten Seam .018"	Sq	1,135.00
Standing Seam .018"	Sq	1,170.00
Flat Seam .108"	Sq	1,080.00
Lead - Copper Coated - Batten Seam 3 lb	Sq	1,070.00
Flat Seam 3 lb	Sq	1,005.00
Aluminum - Clad	Sq	825.00
Galvanized - Colored .020"	Sq	900.00
See Division 5 for Corrugated Type		
.5 EXPANSION JOINTS - 2 1/2"		
Polyethylene with galvanized	LnFt	6.90
Galvanized 25 ga	LnFt	8.03
Aluminum .032	LnFt	9.10
Copper 16 oz	LnFt	22.05
Stainless Steel	LnFt	12.05
Neoprene	LnFt	8.81
Butyl	LnFt	7.05
Add for 3 1/2"	LnFt	10%

		UNIT	COST	or	UNIT	COST
0710.0	**ROOF AND SOFFITT ACCESSORIES (L&M) (Roofers)**					
.1	SKYLIGHTS					
	Example: 4' - 0" x 4' - 0"	SqFt	35.70		Each	600.00
	8' - 0" x 8' - 0"	SqFt	22.40		Each	1,460.00
.2	SKY DOMES (Plastic - Clear or White)					
	20" x 20" Single				Each	165.00
	Double				Each	195.00
	24" x 24" Single				Each	200.00
	Double				Each	230.00
	48" x 48" Single				Each	485.00
	Double				Each	550.00
	72" x 72" Single				Each	815.00
	Double				Each	980.00
	96" x 96" Double				Each	1,550.00
	Add for Bronze Top Colors				-	10%
	Add for Electric Operated Type				Each	4.60
	Add for Pyramid Type				Each	100%
	Add for Hip Type				Each	50%
.3	GRAVITY VENTILATORS					
	Diameter Aluminum or Galvanized 12"				Each	105.00
	24"				Each	250.00
.4	SMOKE VENTILATORS					
	Example: 48" x 96" - Aluminum				Each	2,140.00
	48" x 48" - Steel				Each	2,000.00
.5	PREFAB CURBS & EXPANSION JOINTS - 3" Fibered				LnFt	1.65
	4" Fibered				LnFt	1.70
.6	ROOF LINE LOUVRES				LnFt	1.60
.7	TURBINE VENTILATORS - 12"				Each	325.00
	24"				Each	335.00
.8	CUPOLAS				Each	345.00
.9	ROOF VENTS - P.V.C.				Each	60.00
	Aluminum				Each	55.00
	See 1012 for Roof Hatches					
0711.0	**SEALANTS (L&M) (Carpenters)**					
.1	CAULKING - 1/2" x 1/2" Joints					
	Windows and Doors - 2-Part Proprietary (Oil Base)				LnFt	1.85
	Silicone Rubber				LnFt	2.16
	Polysulfide (Thiakol)				LnFt	2.32
	Acrylic				LnFt	2.26
	Polyurethane				LnFt	2.26
	Control Joints - 2 - Part Proprietary				LnFt	2.04
	Silicone Rubber				LnFt	2.42
	Polysulfide				LnFt	3.30
	Acrylic				LnFt	2.45
	Polyurethane				LnFt	2.45
	Stone Pointing -Silicone Rubber				LnFt	2.80
	Polysulfide				LnFt	3.00
	Acrylic				LnFt	2.42
	Add for Swing Stage Work				LnFt	.45
	Add for 3/4" x 3/4" Joints				LnFt	.90
	Deduct for 1/4" x 1/4" Joints				LnFt	.28
.2	GASKETS AND JOINT FILLERS					
	Gaskets - 1/4" Polyvinyl x 6"				LnFt	2.20
	1/2" Polyvinyl x 6"				LnFt	2.75
	1/4" Neoprene x 6"				LnFt	2.95
	1/2" Neoprene x 6"				LnFt	3.80
	Joint Fillers -1/2"				LnFt	.34
	3/4"				LnFt	.39
	1"				LnFt	.42

	SQ.FT. COST
0701.0 WATERPROOFING	
1-Ply Membrane Felt 15#	1.38
2-Ply Membrane Felt 15#	2.10
Hydrolithic	2.35
Elastomeric Rubberized Asphalt with Poly Sheet	2.15
Liquid	2.35
Metallic Oxide 3-Coat	3.55
Vinyl Plastic	2.40
Bentonite 3/8" - Trowel	1.70
5/8" - Panels	1.90
0702.0 DAMPPROOFING	
Asphalt Trowel Mastic 1/16"	.84
1/8"	1.00
Spray Liquid 1-Coat	.53
2-Coat	.73
Brush Liquid 2-Coat	.80
Hot Mop 1-Coat and Primer	1.40
1 Fibrous Asphalt	1.50
Cementitious Per Coat 1/2" Coat	.68
Silicone 1-Coat	.50
2-Coat	.77
Add for Scaffold and Lift Operations	.45
0703.0 BUILDING INSULATION	
FLEXIBLE	
Fiberglass 2 1/4" R 7.40	.66
3 1/2" R 11.00	.70
3 5/8" R 13.00	.82
6" R 19.00	.92
9" R 30.00	1.20
12" R 38.00	1.30
Add for Ceiling Work	.08
Add for Paper Faced	.09
Add for Polystyrene Barrier (2m)	.08
Add for Scaffold Work	.45
RIGID	
Fiberglass 1" R 4.35 3# Density	.92
1 1/2" R 6.52 3# Density	1.15
2" R 8.70 3# Density	1.39
Styrene, Molded 1" R 4.35	.72
1 1/2" R 6.52	.84
2" R 7.69	1.08
2" T&G R 7.69	1.38
Styrene, Extruded 1" R 5.40	1.01
1 1/2" R 6.52	1.18
2" R 7.69	1.28
2" T&G R 7.69	1.40
Perlite 1" R 2.78	1.00
2" R 5.56	1.58
Urethane 1" R 6.67	1.12
2" R 13.34	1.47
Add for Glued Applications	.08

0703.0 BUILDING INSULATION, Cont'd...	SQ.FT.COST
LOOSE 1" Styrene R 3.8	.70
1" Fiberglass R 2.2	.68
1" Rock Wool R 2.9	.69
1" Cellulose R 3.7	.67
FOAMED Urethane Per Inch	1.73
SPRAYED Cellulose Per Inch	1.26
Polystyrene	1.45
Urethane	2.10
ALUMINUM PAPER	.54
VINYL FACED FIBERGLASS	1.70
0704.0 ROOF & DECK INSULATION - COMBINED WITH 0709.0 BELOW	
0705.0 MEMBRANE ROOFING - COMBINED WITH 0709.0 BELOW	
0708.0 SHINGLE ROOFING	
ASPHALT SHINGLES 235# Seal Down	1.24
300# Laminated	1.53
325" Fire Resistant	1.58
Timberline	1.56
340# Tab Lock	2.04
ASPHALT ROLL 90#	.72
55#	.65
FIBERGLASS SHINGLES 215#	1.24
250#	1.48
RED CEDAR 16" x 5" to Weather #1 Grade	3.30
#2 Grade	2.85
#3 Grade	2.14
24" x 10" to Weather - Hand Splits - 1/2" x 3/4" #2	3.11
3/4" x 1 1/4" #2	3.60
Add for Fire Retardant	.52
METAL Aluminum 020 Mill	3.35
020 Anodized	3.85
Steel, Enameled Colored	4.90
Galvanized Colored	4.00
Add to Above for 15# Felt Underlayment	
Add for Base Starter	.64
Add for Ice & Water Starter	1.00
Add for Boston Ridge	2.20
Add for Removal & Haul Away	.80
Add for Pitches Over 5-12 each Pitch Increase	.06
Add for Chimneys, Skylights and Bay Windows - Each	65.00

0704.0/ 0705.0/ 0709.0 - COMBINED ROOFING INSULATION and SHEET METAL	
MEMBRANE	SQ. COST
3-Ply Asphalt & Gravel - Fiberboard & Perlite - R 10.0	430.00
R 16.6	480.00
4-Ply - R 10.0	490.00
R 16.6	515.00
5-Ply - R 10.0	565.00
R 16.6	625.00
Add for Sheet Rock over Steel Deck - 5/8"	115.00
Add for Fiberglass Insulation	38.00
Add for Pitch and Gravel	58.00
Add for Sloped Roofs	47.00
Add for Thermal Barrier	37.00
Add for Upside Down Roofing System	107.00
SINGLE-PLY - 60M Butylene Roofing & Gravel Ballast - R 10.0	405.00
Mech. Fastened - R 10.0	415.00
PVC & EPDM Roofing & Gravel Ballast - R 10.0	385.00
Mech. Fastened - R 10.0	415.00
Blocking and Cants Not Included	

0801.0	HOLLOW METAL (M) (CARPENTERS)			
.1	CUSTOM	UNIT	LABOR	MATERIAL
.11	Frames (16 Gauge)			
	2'6" x 6'8" x 4 3/4"	Each	51	160
	2'8" x 6'8" x 4 3/4"	Each	52	170
	3'0" x 6'8" x 4 3/4"	Each	53	180
	3'4" x 6'8" x 4 3/4"	Each	62	185
	3'6" x 6'8" x 4 3/4"	Each	70	190
	4'0" x 6'8" x 4 3/4"	Each	80	200
	Deduct for 18 Gauge	Each	8	17.50
	Add for 14 Gauge	Each	12	26
	Add for A, B, or C Label	Each	8	13.25
	Add for Transoms	Each	32	80
	Add for Side Lights	Each	34	98
	Add for Frames over 7'0" (Height)	Each	34	17.50
	Add for Frames over 6 3/4" (Width)	Each	13	17.50
	Add for Bolted Frame with Bolts	Each	24	36
.12	Doors (1 3/4" - 18 Gauge)			
	2'6" x 6'8" x 1 3/4"	Each	54	195
	2'8" x 6'8" x 1 3/4"	Each	55	205
	3'0" x 6'8" x 1 3/4"	Each	57	215
	3'4" x 6'8" x 1 3/4"	Each	65	225
	3'6" x 6'8" x 1 3/4"	Each	74	245
	4'0" x 6'8" x 1 3/4"	Each	85	275
	Add for 16 Gauge	Each	11	41
	Deduct for 20 Gauge	Each	8	17.50
	Add for A Label	Each	9	17.50
	Add for B or C Label	Each	8	17.50
	Add for Vision Panels or Lights	Each	-	77
	Add for Louvre Openings	Each	-	72
	Add for Galvanizing	Each	-	47
.2	STOCK			
.21	Frames (4 3/4"-5 3/4"-6 3/4"-8 3/4"- 16 Ga)			
	2'6" x 6'8" or 7'0"	Each	50	95
	2'8" x 6'8" or 7'0"	Each	51	103
	3'0" x 6'8" or 7'0"	Each	54	108
	3'4" x 6'8" or 7'0"	Each	60	113
	3'6" x 6'8" or 7'0"	Each	68	120
	4'0" x 6'8" or 7'0"	Each	77	145
	See above .11 for Size Change			
	Add for Bolted Frame & Bolts	Each	21	36
	Add for A, B, or C Label	Each	8	17.50
	Deduct Non-Welded Frame (Knock Down)	Each	6	17.50
.22	Doors (1 3/8" or 1 3/4" - 18 Gauge)			
	2'6" x 6'8" or 7'0"	Each	54	200
	2'8" x 6'8" or 7'0"	Each	55	205
	3'0" x 6'8" or 7'0"	Each	57	215
	3'4" x 6'8" or 7'0"	Each	64	225
	3'6" x 6'8" or 7'0"	Each	72	260
	4'0" x 6'8" or 7'0"	Each	79	270
	Add for A Label	Each	8	17.50
	Add for B or C Label	Each	7	12.25
	Add for Galvanizing	Each	-	47
	Labor Above Includes Installation of Butts & Locksets			
	Add to All Above for:			
	Door Closures - Surface Mounted	Each	52	107
	Panic Bars (Exit Device)	Each	105	495
	Kick, Push and Pull Plates	Each	21	35

See 8-13 for Other Hardware

0802.0 WOOD DOORS (M) (CARPENTERS)

See 0603.3 and 0802.5 for Custom Frames
Labor includes Installation of Butts and Locksets

.1 FLUSH DOORS - Pre-Machined	UNIT	LABOR	Hollow Core	Solid Core
No Label - Standard - Birch - 7 Ply - Paint Grade				
2'0" x 6'8" or 1 3/8"	Each	52	105	115
2'4" x 6'8" or 1 3/8"	Each	53	110	120
2'6" x 6'8" or 1 3/8"	Each	54	115	125
2'8" x 6'8" or 1 3/8"	Each	55	120	130
3'0" x 6'8" or 1 3/8"	Each	57	125	140
2'0" x 6'8" or 1 3/4"	Each	53	110	125
2'4" x 6'8" or 1 3/4"	Each	54	115	135
2'6" x 6'8" or 1 3/4"	Each	55	120	140
2'8" x 6'8" or 1 3/4"	Each	56	120	145
3'0" x 6'8" or 1 3/4"	Each	60	125	150
3'4" x 6'8" or 1 3/4"	Each	64	145	170
3'6" x 6'8" or 1 3/4"	Each	68	220	250
3'8" x 6'8" or 1 3/4"	Each	74	240	260
Add for Stain Grade	Each			40
Add for 5 Ply	Each			80
Add for Jamb & Trim - Solid - Knock Down	Each			80
Add for Jamb & Trim- Veneer- Knock Down	Each			70
Add for Red Oak - Rotary	Each			23
Add for Red Oak - Plain Sliced	Each			40
Add for Architectural Grade - 7 Ply	Each			85
Deduct for Lauan	Each			15
Add for Doors over 6'8" per Inch	Each			15
Add for Vinyl Overlay	Each			47
Add for Cutouts and Special Cuts	Each			54
Add for Lite Cutouts with Metal Frames	Each			79
Add for Private Door Eye	Each			22
Add for Wood Louvre	Each			110
Add for Transom Panels & Side Panels	Each	26		120
Add for Astragals	Each	27		52
Add for Trimming Door for Carpet	Each	23		-

			MATERIAL		
Label - Birch - Paint Grade	UNIT	LABOR	20-Min	45-Min	60-Min
2'0" x 6'8" x 1 3/4"	Each	52	140	170	180
2'4" x 6'8" x 1 3/4"	Each	53	150	180	195
2'6" x 6'8" x 1 3/4"	Each	54	155	190	200
2'8" x 6'8" x 1 3/4"	Each	55	160	235	250
3'0" x 6'8" x 1 3/4"	Each	56	165	275	280
3'4" x 6'8" x 1 3/4"	Each	63	180	295	290
3'6" x 6'8" x 1 3/4"	Each	68	260	310	310
3'8" x 6'8" x 1 3/4"	Each	75	285	320	330
See Above 0602.1 for Changes					

			MATERIAL		
.2 PANEL DOORS	UNIT	LABOR	Pine	Fir	Oak
Exterior Entrances					
2'8" x 6'8" x 1 3/4"	Each	80	350	360	560
3'0" x 6'8" x 1 3/4"	Each	85	370	380	580
Add for Sidelights	Each	40	205	225	360
Interior					
2'6" x 6'8" x 1 3/8"	Each	60	310	290	460
2'8" x 6'8" x 1 3/8"	Each	65	320	300	470
3'0" x 6'8" x 1 3/8"	Each	70	350	310	480

0802.0 WOOD DOORS (Cont'd...)

		UNIT	LABOR	MATERIAL Birch/Oak	Pine	Lauan
.3	LOUVERED DOORS					
	1'0" x 6'8" x 1 3/8" Pine	Each	35	400	150	-
	1'3" x 6'8" x 1 3/8"	Each	38	430	155	-
	1'6" x 6'8" x 1 3/8"	Each	42	440	160	-
	2'0" x 6'8" x 1 3/8"	Each	51	520	185	-
	2'4" x 6'8" x 1 3/8"	Each	56	520	200	-
	2'6" x 6'8" x 1 3/8"	Each	56	530	250	-
	2'8" x 6'8" x 1 3/8"	Each	58	550	260	-
	3'0" x 6'8" x 1 3/8"	Each	70	565	290	-
	Add for Half Louvered/Panel	Each				15%
	Add for Prehung	Each	-	95	95	0
.4	BI-FOLD - FLUSH					
	2'0" x 6'8" x 1 3/8" Pine	Each	51	-	-	88
	2'4" x 6'8" x 1 3/8"	Each	54	-	-	95
	2'6" x 6'8" x 1 3/8"	Each	58	-	-	95
	2'8" x 6'8" x 1 3/8"	Each	60	-	-	100
	3'0" x 6'8" x 1 3/8"	Each	65	-	-	108
	Add for Oak	Each	-	39	-	-
	Add for Birch	Each	-	34	-	-
	Add for Louvered Type	Each	10			85
	Add for Half Louvered/Panel	Each	10			85
	Add for Decorative Type	Each	10			175
	Add for Prefinished Type	Each	-			22
	Add for Heights over 6'8"	Each	10			5
	Add for Widths over 3'0"	Each	10			6
	See 0803.2 for Closet Bi-fold Doors and Prehung Stock					
.5	CAFE DOORS			Oak	Pine	Fir
	2'6" x 3'8" x 1 1/8"	Pair	68	-	250	260
	2'8" x 3'8" x 1 1/8"	Pair	75	-	270	270
	3'0" x 3'8" x 1 1/8"	Pair	80	-	280	280
.6	FRENCH - with Grille Pattern					
	2'6" x 6'8" x 1 3/8"	Each	80	500	570	350
	2'8" x 6'8" x 1 3/8"	Each	82	510	580	380
	3'0" x 6'8" x 1 3/8"	Each	85	520	590	520
	Add for Prehung	Each	-	95	-	-
.7	DUTCH					
	2'6" x 6'8" x 1 3/4"	Each	102	-	700	-
	2'8" x 6'8" x 1 3/4"	Each	107	-	775	-
	3'0" x 6'8" x 1 3/4"	Each	112	-	800	-
.8	PREHUNG DOOR UNITS - Incl. Frame, Trim, Hardware and Jambs					
	Exterior			Birch	Pine	Fir
	2'8" x 6'8" x 1 3/4"	Each	68	-	320	-
	3'0" x 6'8" x 1 3/4"	Each	75	-	370	-
	3'4" x 6'8" x 1 3/4"	Each	80	-	470	-
	Add for Insulated	Each	10	-	-	140
	Add for Threshold	Each	10	-	-	18
	Add for Weatherstrip	Each	-	-	-	87
	Add - Sidelites 1'0" HC	Each	24	-	-	165
	Add - Sidelites 1'6" HC	Each	28	-	-	200
	Interior			Birch	Lauan	Oak
	2'6" x 6'8" x 1 3/8"	Each	63	215	150	220
	2'8" x 6'8" x 1 3/8"	Each	67	225	155	225
	3'0" x 6'8" x 1 3/8"	Each	74	235	160	235

0803.0 SPECIAL DOORS

				MATERIAL			
.1	BLAST OR NUCLEAR RESISTANT						
.2	CLOSET - BI-FOLDING (PREHUNG STOCK)						
.21	Wood - Units - 1 3/8"	UNIT	LABOR	Oak Flush	Birch Flush	Lauan Flush	Pine & Fir Panel
	Incl. Jamb, Csg. Track & Hdwe.						
	2 Door - 2'0" x 6'8"	Each	46	175	190	125	420
	2'6" x 6'8"	Each	50	182	200	130	450
	3'0" x 6'8"	Each	54	190	210	137	500
	4 Door - 4'0" x 6'8"	Each	60	285	280	195	700
	5'0" x 6'8"	Each	67	295	300	200	800
	6'0" x 6'8"	Each	75	315	325	225	865
	Incl. Track & Hdwe (Unfinished)						
	2 Door - 2'0" x 6'8"	Each	39	105	98	98	-
	2'6" x 6'8"	Each	44	115	104	105	-
	3'0" x 6'8"	Each	46	125	115	115	-
	4 Door - 4'0" x 6'8"	Each	49	170	150	150	-
	5'0" x 6'8"	Each	52	183	180	165	-
	6'0" x 6'8"	Each	56	205	200	180	-
	Add for Louvered Type	Each					100%

		UNIT	LABOR	MATERIAL
.22	Metal			
	Flush 2 Door - 2'0" x 6'8"	Each	38	90
	2'6" x 6'8"	Each	40	95
	3'0" x 6'8"	Each	42	100
	Flush 4 Door - 4'0" x 6'8"	Each	46	125
	5'0" x 6'8"	Each	48	130
	6'0" x 6'8"	Each	54	140
	Add for Plastic Overlay	Each	-	32
	Add for Louvered or Decorative	Each	-	10
.23	Leaded Mirror			
	4 Panel Unit - 4'0" x 6'8"	Each	84	595
	5'0" x 6'8"	Each	90	650
	6'0" x 6'8"	Each	100	700
.3	FLEXIBLE DOORS (M) (CARPENTER)			
	(See 1020 - Retractable Partitions)			
	Wood Slat (Unfinished)	SqFt	2.00	11.50
	Fabric Accordion Fold	SqFt	2.00	9.50
	Vinyl Clad	SqFt	2.00	13.00
.4	GLASS-ALL (L&M) (GLAZIERS)			COST
	Manual (Including Hardware)	Each	-	2,200
	Add for Center Locking	Each	-	175
	Automatic (Including Hardware & Operator)			
	Mat or Floor Unit	Each	-	6,300
	Handle Unit	Each	-	6,800
.5	HANGAR (L&M) (IRONWORKERS)			
	Biparting or Sliding - Steel			
	Example: 100' x 20'	Each	-	55,000
	Add for Pass Doors	Each	-	1,350
	Add for Electric Operator	Each	-	1,750
	Add for Insulating	Each	-	1,950
	Tilt Up or Canopy	Each	Approx. Same	
.6	METAL COVERED FIRE & INDUSTRIAL SLIDING DOORS			
	3'0" x 7'0" - Flush	Each	-	680
	4'0" x 7'0"	Each	-	1,950
	8'0" x 8'0" - U.L. Label	Each	-	2,600
	10'0" x 10'0"	Each	-	3,300
	10'0" x 12'0"	Each	-	3,600
	12'0" x 12'0"	Each	-	4,000

0803.0 SPECIAL DOORS, Cont'd...

.7 OVERHEAD DOORS (L&M) (CARPENTERS)

	UNIT	COST 1 3/4" Wood	 2" 24-Ga Steel
Commercial			
8' x 8'	Each	850.00	860.00
8' x 10'	Each	1,000.00	1,023.00
8' x 12'	Each	1,100.00	1,150.00
10' x 10'	Each	1,250.00	1,350.00
10' x 12'	Each	1,350.00	1,400.00
12' x 12'	Each	1,500.00	1,550.00
14' x 10'	Each	1,550.00	1,600.00
14' x 12'	Each	1,900.00	1,950.00
Add for Panel Door	Each		5%
Add for Low Lift	Each		78.00
Add for High Lift	Each		130.00
Add for Electric Operator	Each		575.00
Add for Insulated Type	Each		10%
Residential			
8' x 7' x 1 3/4"	Each	470.00	480.00
9' x 7' x 1 3/4"	Each	500.00	500.00
16' x 7' x 1 3/4"	Each	1,020.00	1,100.00
Add for Wood Paneled Door	Each		10%
Add for Redwood	Each		225.00
Add for Electric Operator	Each		425.00
Add for Vision Lights	Each		22.00

.8 PLASTIC LAMINATE FACED (M) (CARPENTERS)

	UNIT	LABOR	MATERIAL
1 3/8"	SqFt	4.30	12.00
1 3/4"	SqFt	4.40	12.00
Add for Edge Strip - Vertical	Each		7.25
Add for Edge Strip - Horizontal	Each		6.30
Add for Cutouts	Each		60.00
Add for Metal Frame Cutout	Each		87.00
Add for Decorative Type	SqFt		19.00
Add for Solid Colors	SqFt		10%
Add for 1-hour B Label	SqFt		5.10
Add for 1 1/2-hour B Label	SqFt		5.25
Add for Color Core	-		50%
Add for Machining for Hardware	Each		50.00

.9 REVOLVING DOOR (L&M) (GLAZIERS)

		COST
Aluminum		
3 Leaf - 6' - 6" x 6' - 6"	Each	22,500.00
4 Leaf - 6' - 6" x 6' - 6"	Each	24,000.00
6' - 0" x 7' - 0"	Each	17,500.00
Automatic - 6' - 6" x 10' - 0"	Each	34,000.00
Stainless	Each	29,500.00
Bronze	Each	33,000.00
Painted Steel - Dark Room 32"	Each	1,950.00
36"	Each	2,000.00

0803.0	SPECIAL DOORS, Cont'd...	UNIT	LABOR	MATERIAL
.10	ROLLING DOORS & GRILLES (M) (IRONWORKERS)			
	Doors - Manual - 8' x 8'	Each	410.00	1,950.00
	10' x 10'	Each	485.00	2,000.00
	12' x 12'	Each	475.00	2,700.00
	14' x 14'	Each	740.00	3,800.00
	Add for Electric Controlled	Each	-	300.00
	Add for Fire Doors	Each	-	660.00
	Grilles - Manual -6'8" x 3'2"	Each	180.00	800.00
	6'8" x 4'2"	Each	165.00	890.00
	8'0" x 3'2"	Each	200.00	940.00
	8'0" x 4'2"	Each	210.00	1,050.00
	Add for Electric Controlled	Each	-	280.00
.11	SHOWER DOORS (M) (CARPENTERS)			
	28" x 66" Obscure Pattern Glass Wire	Each	57.00	150.00
	Tempered Glass	Each	57.00	160.00
.12	SLIDING OR PATIO DOORS (M) (CARPENTERS)			
	ONE SLIDING, ONE FIXED			
	Metal (Aluminum) (Including 5/8" Glass, Threshold, Hardware and Screen)			
	8'0" x 6'10"	Each	150.00	1,175.00
	8'0" x 8'0"	Each	200.00	1,300.00
	Wood (M) (Carpenters) Vinyl Sheathed (Including 5/8" Insulated Dbl. Temp. Glass, Weatherstripped, Hardware and Casing)			
	Vinyl Clad 6' 0" x 6' 8"	Each	150.00	1,600.00
	8' 0" x 6' 8"	Each	180.00	1,900.00
	Pine - Prefinished 6' 0" x 6' 8"	Each	150.00	1,600.00
	8' 0" x 6' 8"	Each	178.00	1,800.00
	Deduct for 3/16" Safety Glass/Insulated	Each	-	175.00
	Add for Tripe Glazing	Each	-	210.00
	Add for Screen	Each	16.00	145.00
	Add for Grilles	Each	20.00	220.00
.13	SOUND REDUCTION (M) (CARPENTERS)			
	Metal	Each	165.00	1,700.00
	Wood	Each	154.00	900.00
.14	TRAFFIC (M) (CARPENTERS)			
	<u>Electric Powered</u>			
	Hollow Metal - 8' x 8'	Each	415.00	4,700.00
	8' x 8' High Speed	Each	1,050.00	10,500.00
	Metal Clad (26 ga.) - 8' x 8'	Each	390.00	4,500.00
	Wood	Each	340.00	3,700.00
	<u>Truck Impact - Double Acting</u>			
	(Including Bumpers and Vision Panels)			
	Metal Clad - 8' x 8' Openings	Each	295.00	2,600.00
	Rubber Plastic - 8' x 8' Openings 1/2"	Each	235.00	2,150.00
	Aluminum - 8' x 8' Openings 1 3/4"	Each	255.00	2,700.00
.15	SPECIAL MADE & ENGINEERED INDUSTRIAL DOORS (L&M) (CARPENTERS)			
	Four-Fold Garage Doors - Electric Operated	SqFt	-	105.00
.16	VAULT DOORS (M) (STEEL ERECTORS)			
	1 hour with Frame 6' 6" x 2' 8"	Each	345.00	2,200.00
	2 hour with Frame 6' 6" x 2' 8"	Each	345.00	2,900.00
	6' 6" x 3' 0"	Each	415.00	3,300.00
	6' 6" x 3' 4"	Each	465.00	3,600.00

0804.0 ENTRANCE DOORS, FRAMES & STORE FRONT CONSTRUCTION

Prices Include Hardware & Safety Plate Glazing

		UNIT	COST
.1	ALUMINUM		
.11	Stock		
	Door and Frame - 3' x 7'	Each	1,500.00
	6' x 7'	Each	2,750.00
	Window or Store Front Framing - 10'	SqFt	27.00
.12	Custom		
	Door and Frame - 3' x 7'	Each	1,800.00
	6' x 7'	Each	3,450.00
	Window or Store Front Framing - 10'	SqFt	34.00
	Add for Anodized Finish		15%
.2	BRONZE		
	Door and Frame - 3' x 7'	Each	3,800.00
	Window or Store Front Framing	SqFt	66.00
.3	STAINLESS STEEL		
	Door and Frame - 3' x 7'	Each	3,400.00
	Window or Store Front Framing	SqFt	68.00
	Add for Insulated Glass	SqFt	6.60
	Add for Exit Hardware	Each	350.00

See 0803.10 for Revolving Doors
See 0803.0 for Automatic Door Opening Assemblies

0805.0 METAL WINDOWS (M) (CARPENTERS OR STEEL ERECTORS) (SINGLE GLAZING INCLUDED)

All Example Sizes Below: 2'4" x 4'6"

		UNIT	LABOR	MATERIAL	or	UNIT	LABOR	MATERIAL
.1	ALUMINUM WINDOWS							
	Casement and Awning	Each	52.00	250.00		SqFt	5.00	23.90
	Sliding or Horizontal Rolling	Each	52.00	180.00		SqFt	5.00	18.00
	Double and Single-Hung or Vertical Sliding	Each	53.00	180.00		SqFt	5.10	18.00
	Projected	Each	49.00	230.00		SqFt	4.75	23.00
	Add for Screens					SqFt	.67	4.00
	Add for Storms					SqFt	1.40	6.50
	Add for Insulated Glass					SqFt	-	6.75
	Add for Bronzed Finish					SqFt	-	10%
.2	ALUMINUM SASH							
	Casement	Each	42.00	230.00		SqFt	3.95	23.00
	Sliding	Each	42.00	160.00		SqFt	3.95	16.00
	Single-Hung	Each	42.00	160.00		SqFt	3.95	16.00
	Projected	Each	42.00	250.00		SqFt	3.95	25.00
	Fixed	Each	42.00	130.00		SqFt	3.95	13.00
.3	STEEL WINDOWS							
	Double-Hung	Each	49.00	365.00		SqFt	4.75	36.00
	Projected	Each	49.00	305.00		SqFt	4.75	30.50
	Add for Insulated Glass					SqFt	-	5.75
.4	STEEL SASH							
	Casement	Each	42.00	260.00		SqFt	3.95	26.00
	Double-Hung	Each	42.00	345.00		SqFt	3.95	36.00
	Projected	Each	42.00	295.00		SqFt	3.95	31.00
	Fixed	Each	42.00	170.00		SqFt	3.95	18.00

0806.0 WOOD WINDOWS (M) (Carpenters)

All Units Assembled, Kiln Dry Pine, Glazed, Vinyl Clad, Weather-stripped, and No Interior Trim. Top Quality. Other Sizes Available. Most commonly used listed below. Deduct 10% for Medium Quality and 25% for Low Quality Woods.

	UNIT	LABOR	MATERIAL
.1 BASEMENT OR UTILITY			
Prefinished 2'8" x 1'4"	Each	40.00	120.00
2'8" x 1'8"	Each	47.00	128.00
2'8" x 2'0"	Each	49.00	138.00
Add for Double Glazing	Each	-	40.00
Add for Storm	Each	15.50	28.00
Add for Screen	Each	12.50	18.00
.2 CASEMENT OR AWNING - Operable Units			
Single Clear Tempered Glass 1'8" x 3'0"	Each	51.00	300.00
1'8" x 3'6"	Each	52.00	310.00
2'4" x 3'6"	Each	59.00	330.00
2'4" x 4'0"	Each	61.00	395.00
2'4" x 5'0"	Each	68.00	440.00
Double (2 Units) 2'10" x 3'0"	Each	74.00	500.00
3'4" x 3'6"	Each	79.00	620.00
4'0" x 3'0"	Each	74.00	520.00
4'0" x 3'6"	Each	80.00	640.00
4'0" x 5'0"	Each	84.00	770.00
4'0" x 5'8"	Each	88.00	840.00
Triple (3 Units, 1 Fixed) 6'0" x 3'0"	Each	85.00	760.00
6'0" x 4'0"	Each	90.00	880.00
6'0" x 4'6"	Each	100.00	1,020.00
Add for Grille Patterns - Interior - Polycarbonate	Each	-	22.00
Add for Grille Patterns - Exterior - Prefinished	Each	-	60.00
Add for High Performance Glazing	Each	-	53.00
Add for Screen - per Unit	Each	-	19.00
Add for Extension Jambs	Each	-	28.00
Add for Blinds	Each	-	90.00
CASEMENT OR AWNING - Stationary & Fixed Units			
Single 1'8" x 3'0"	Each	51.00	250.00
1'8" x 3'6"	Each	52.00	280.00
2'4" x 3'6"	Each	59.00	290.00
2'4" x 4'0"	Each	62.00	360.00
2'4" x 5'0"	Each	69.00	390.00
2'4" x 6'0"	Each	73.00	450.00
Picture 4'0" x 4'0"	Each	44.00	465.00
4'0" x 4'6"	Each	48.00	495.00
4'0" x 6'0"	Each	50.00	660.00
6'0"x 4'0"	Each	51.00	650.00
6'0" x 6'0"	Each	56.00	860.00
Add for Grille Patterns - Wood - 2'4"	Each	23.00	85.00
Add for Grille Patterns - Wood - 4'6"	Each	28.00	135.00
.3 DOUBLE HUNG (Insulating Glass with Screen)			
2'6" x 3'0"	Each	42.00	320.00
2'6" x 3'6"	Each	45.00	330.00
2'6" x 4'2"	Each	48.00	350.00
3'2" x 3'0"	Each	54.00	340.00
3'2" x 3'6"	Each	56.00	370.00
3'2" x 4'2"	Each	59.00	385.00
Deduct for Primed Units	Each	5.00	27.00
Add for Grille Patterns	Each	8.00	27.00
Add for Triple Glazing	Each	-	63.00
Deduct for Screen Unit	Each	10.00	20.00

0806.0	WOOD WINDOWS, Cont'd...	UNIT	LABOR	MATERIAL
.4	GLIDER WITH SCREEN & INSUL GLASS 4'0" x 3'6"	Each	54.00	870.00
	5'0" x 3'6"	Each	62.00	925.00
	4'0" x 4'0"	Each	73.00	900.00
	5'0" x 4'0"	Each	78.00	975.00
.5	PICTURE WINDOW WITH CASEMENTS (2)			
	& INSULATING GLASS 9'6" x 5'0"	Each	140.00	1,800.00
	10'1" x 5'0"	Each	145.00	1,900.00
	9'6" x 5'6"	Each	140.00	1,900.00
	SINGLE GLASS AND STORM 10'3" x 4'6"	Each	125.00	1,925.00
	10'3" x 5'6"	Each	140.00	1,975.00
	10'3" x 6'6"	Each	145.00	2,050.00
	Add for High Performance Glass	Each	-	160.00
.6	CASEMENT ANGLE BAY WINDOWS, INSULATING GLASS, VINYL SHEATHED			
	30° (3 units) 5'10" x 4'2" x 2 7/8"	Each	195.00	1,450.00
	(4 units) 7'10" x 4'2" x 2 7/8"	Each	200.00	1,750.00
	(3 units) 5'10" x 5'2" x 2 7/8"	Each	190.00	1,570.00
	(4 units) 7'10" x 5'2" x 2 7/8"	Each	200.00	1,750.00
	(5 units) 9'10" x 4'2" x 2 7/8"	Each	235.00	2,050.00
	(5 units) 9'10" x 5'2" x 2 7/8"	Each	245.00	2,250.00
	45° (3 units) 5'4" x 4'2" x 2 7/8"	Each	200.00	1,450.00
	(4 units) 7'4" x 4'2" x 2 7/8"	Each	198.00	1,580.00
	(3 units) 5'4" x 5'2" x 2 7/8"	Each	200.00	1,580.00
	(4 units) 7'4" x 5'2" x 2 7/8"	Each	210.00	1,750.00
	(5 units) 9'4" x 4'2" x 2 7/8"	Each	225.00	2,150.00
	(5 units) 9'4" x 5'2" x 2 7/8"	Each	245.00	2,350.00
	Add for Screens	Each	47.00	50.00
	Add for Combination Windows	Each	-	245.00
	Deduct for Triple Glazing	Each	-	255.00
	Add for High Performance Glazing	Each	-	68.00
	Add for Bronze Glazing	Each	-	148.00
	Add for Color	Each	-	69.00
	Add for Blinds	Each	-	360.00
	Add for Bottom or Roof Skirt - Plywood	Each	94.00	400.00
	Add for Bottom or Roof Skirt - Aluminum	Each	100.00	575.00
.7	CASEMENT BOW WINDOWS, INSULATING GLASS			
	(3 units) 6'2" x 4'2"	Each	295.00	1,250.00
	(4 units) 8'2" x 4'2"	Each	355.00	1,680.00
	(3 units) 6'2" x 5'2"	Each	305.00	1,560.00
	(4 units) 8'2" x 5'2"	Each	380.00	1,950.00
.8	90° CASEMENT BOX BAY WINDOWS, INSULATING GLASS			
	4'8" x 4'2"	Each	385.00	1,780.00
	6'8" x 4'2"	Each	468.00	2,250.00
	4'8" x 5'2"	Each	395.00	1,810.00
	6'8" x 5'2"	Each	500.00	2,350.00
	Add for Screens - per Unit	Each	47.00	69.00
	Deduct for Double Glazing - per Unit	Each	-	37.00
	Add for High Performance Glazing	Each	-	74.00

		UNIT	LABOR	STATIONARY	VENT
.9	ROOF WINDOWS - INSULATING GLASS				
	1'10" x 3'10"	Each	183.00	410.00	670.00
	2'4" x 3'10"	Each	190.00	500.00	745.00
	3'8" x 3'10"	Each	208.00	600.00	900.00
	Add for Shingle Flashing	Each	41.00	78.00	74.00
	Add for High Performance Glass	Each	-	83.00	80.00

		UNIT	LABOR	MATERIAL
.10	CIRCLE TOPS - 2'4"	Each	107.00	525.00
	4'0"	Each	117.00	680.00
	4'8"	Each	123.00	800.00
	6'0"	Each	134.00	1,300.00

		UNIT	LABOR	MATERIAL
0807.0	**SPECIAL WINDOWS (M) (Carpenters/ Steel Workers)**			
.1	LIGHT-PROOF WINDOWS	SqFt	5.35	30.00
.2	PASS WINDOWS	SqFt	4.60	23.00
.3	DETENTION WINDOWS			
	Aluminum	SqFt	4.80	33.00
	Steel	SqFt	4.90	31.00
.4	VENETIAN BLIND WINDOWS (ALUMINUM)	SqFt	4.50	26.00
.5	SOUND-CONTROL WINDOWS	Each	4.20	20.00
0808.0	**DOOR AND WINDOW ACCESSORIES**			
.1	STORMS AND SCREENS (CARPENTERS)			
	Windows			
	Screen Only - Wood - 3' x 5'	Each	16.80	78.00
	Aluminum - 3' x 5'	Each	17.80	89.00
	Storm & Screen Combination - Aluminum	Each	20.00	108.00
	Doors			
	Screen Only - Wood - 3' x 6' - 8'	Each	29.00	190.00
	Aluminum	Each	27.00	200.00
	Storm & Screen Combination - Aluminum	Each	32.00	300.00
	Wood 1 1/8"	Each	34.00	285.00
.2	DETENTION SCREENS (M) (CARPENTERS)			
	4' 0" x 7' 0"	Each	84.00	525.00
.3	DOOR OPENING ASSEMBLIES (L&M) (GLAZIERS)			
	Floor or Overhead Electric Eye Units			
	Swing - Single 3' x 7' Door - Hydraulic	Each	-	4,000.00
	Double 6' x 7' Door - Hydraulic	Each	-	5,500.00
	Sliding - Single 3' x 7' Door - Hydraulic	Each	-	5,000.00
	Double 5' x 7' Door - Hydraulic	Each	-	7,000.00
	Industrial Door - 10' x 8	Each	-	6,800.00
.4	SHUTTERS - FOLDING (CARPENTERS)			
	16" x 1 1/8" x 48"	Pair	16.80	95.00
	x 60"	Pair	19.00	110.00
	x 72"	Pair	23.00	120.00

		UNIT	COST
0809.0	**CURTAIN WALL SYSTEMS (L&M) (IRONWORKERS)**		
.1	STRUCTURAL OR VERTICAL TYPE		
	Aluminum Tube Frame, Panel, Sills and Mullions - No glazing		
	Example: 4' Panel, 7' Fixed	SqFt	44.00 to 135.00
.2	PANEL WALL OR HORIZONTAL TYPE		
	Aluminum Tube Frame, Panel, Sills and Mullions - No glazing		
	Example: 1' Panel, 2' Oper., 4' Fixed	SqFt	40.00 to 110.00
.3	ARCHITECTURAL PANELS, INSULATED		
	Included in Above Prices		
	Metal		
	Baked Enamel on Steel - 24 ga	SqFt	14.60
	on Aluminum - 24 ga	SqFt	15.00
	Porcelain Enamel on Steel - 24 ga	SqFt	16.50
	on Aluminum - 24 ga	SqFt	16.90
	Exposed Aggregate - Resin	SqFt	12.80
	Fiberglass	SqFt	17.00
	Spandrel Glass	SqFt	15.00

See 0812 for Glazing Costs to be Added to Above

0810.0 FINISH HARDWARE (M) (CARPENTERS)

		EACH	LABOR	MATERIAL Painted	Bronze	Chrome
.1	BUTTS					
	3" x 3"	Pair	15.50	9.00	10.00	13.00
	3 1/2" x 3 1/2"	Pair	16.00	10.50	10.00	14.50
	4" x 4"	Pair	17.50	13.50	13.00	15.00
	4 1/2" x 4 1/2"	Pair	18.00	18.00	18.00	25.00
	4" x 4" Ball Bearing	Pair	19.00	34.00	34.00	43.00
	4 1/2" x 4 1/2" Ball Bearing	Pair	20.00	39.00	39.00	45.00

		UNIT	LABOR	MATERIAL
.2	CATCHES - ROLLER	Each	15.50	16.00
.3	CLOSURES - SURFACE MOUNTED - 3' - 0" Door	Each	47.00	105.00
	3' - 4"	Each	48.00	108.00
	3' - 8"	Each	49.00	110.00
	4' - 0"	Each	55.00	120.00
	CLOSURES - CONCEALED OVERHEAD - Interior	Each	79.00	165.00
	Exterior	Each	84.00	275.00
	Add for Fusible Link - Electric	Each	21.00	170.00
	CLOSURES - FLOOR HINGES - Interior	Each	115.00	525.00
	Exterior	Each	115.00	550.00
	Add for Hold Open Feature	Each	-	76.00
	Add for Double Acting Feature	Each	-	250.00
.4	DEAD BOLT LOCK - Cylinder - Outside Key	Each	37.00	130.00
	Cylinder - Double Key	Each	42.00	145.00
	Flush - Push/ Pull	Each	23.00	24.00
.5	EXIT DEVICES (PANIC) - Surface	Each	103.00	520.00
	Mortise Lock	Each	107.00	620.00
	Concealed	Each	150.00	1,900.00
	(Hand drop) Automatic	Each	117.00	1,400.00
.6	HINGES, SPRING (PAINTED) - 6" Single Acting	Each	28.00	65.00
	6" Double Acting	Each	28.00	83.00
.7	LATCHSETS - Bronze or Chrome	Each	38.00	135.00
	Stainless Steel	Each	38.00	155.00
.8	LOCKSETS - Mortise - Bronze or Chrome - H.D.	Each	35.00	200.00
	S.D.	Each	33.00	145.00
	Stainless Steel	Each	33.00	270.00
	Cylindrical - Bronze or Chrome	Each	37.00	160.00
	Stainless Steel	Each	37.00	205.00
.9	LEVER HANDICAP - Latch Set	Each	35.00	170.00
	Lock Set	Each	35.00	210.00
.10	PLATES - Kick - 8" x 34" - Aluminum	Each	21.00	21.00
	Bronze	Each	22.00	44.00
	Push - 6" x 15"- Aluminum	Each	20.00	22.00
	Bronze	Each	21.00	36.00
	Push & Pull Combination - Aluminum	Each	23.00	50.00
	Bronze	Each	23.00	83.00
.11	STOPS AND HOLDERS			
	Holder - Magnetic (No Electric)	Each	45.00	125.00
	Bumper	Each	26.00	19.00
	Overhead - Bronze, Chrome or Aluminum	Each	26.00	68.00
	Wall Stops	Each	21.00	15.00
	Floor Stops	Each	21.00	16.00

See 0808.3 for Automatic Openers & Operators

0811.0 WEATHERSTRIPPING (L&M) (CARPENTERS)

		UNIT	LABOR	MATERIAL
.1	ASTRAGALS - Aluminum - 1/8" x 2"	Each	21.00	14.00
	Painted Steel	Each	21.00	14.00
.2	DOORS (WOOD) - Interlocking	Each	35.00	32.00
	Spring Bronze	Each	30.00	52.00
	Add for Metal Doors	Each	30.00	9.00
.3	SWEEPS - 36" WOOD DOORS - Aluminum	Each	16.00	15.00
	Vinyl	Each	16.00	10.00
.4	THRESHOLDS - Aluminum - 4" x 1/2"	Each	20.00	17.00
	Bronze - 4" x 1/2"	Each	21.00	32.00
	5 1/2" x 1/2"	Each	22.00	37.00
.5	WINDOWS (WOOD) - Interlocking	Each	38.00	26.00
	Spring Bronze	Each	33.00	47.00

0812.0	GLASS AND GLAZING (L&M) (Glaziers)	UNIT	COST
.1	BEVELED GLASS - 1/4" x 1/2" Bevel	SqFt	120.00
.2	COATED - Used with Insulated Glass and Heat Absorber and Light Reflector	SqFt	28.00
.3	HEAT ABSORBING AND SOLAR REJECTING COMBINATIONS of Tinted, Coated and Clear Glass in Insulated Glass Form with 1/4" or 1/2" Air Space. Usual thickness with 1/4" air space, 9/16", 11/16" and 1 3/16". With 1/2" air space, 13/16", 15/16" and 1 /16". Factory priced. Greatly variable.	SqFt	21.00
.4	INSULATED GLASS		
	1/2" and 5/8" Welded	SqFt	15.30
	1" Clear	SqFt	17.00
	Tempered Insulated	SqFt	27.00
	Triple Insulated	SqFt	24.00
	Add for Tinting	SqFt	10%
.5	LAMINATED (SAFETY AND SOUND)		
	7/32"	SqFt	14.50
	1/4"	SqFt	17.30
	3/8"	SqFt	18.40
	1/2"	SqFt	26.50
	1 3/16", 1 1/2", 1 3/4" and 2" Bulletproof	SqFt	55.00
.6	MIRROR GLASS		
	Copper Plated Back and Polished Edges	SqFt	13.40
	Unfinished	SqFt	8.25
.7	PATTERNED OR ROUGH (OBSCURE)		
	1/8"	SqFt	6.70
	7/32"	SqFt	6.90
.8	PLASTIC GLASS (SAFETY)		
	Acrylic - 1/8" .125"	SqFt	11.50
	3/16" .187"	SqFt	14.00
	1/4" .250"	SqFt	14.60
	5/16" .312"	SqFt	16.80
	3/8" .375"	SqFt	21.00
	Polycarbonates - Add to Above	SqFt	50%
.9	PLATE - CLEAR		
	1/4" (See .4 for Tinted)	SqFt	7.50
	3/8"	SqFt	10.30
	1/2"	SqFt	16.00
	5/8"	SqFt	18.80
	3/4"	SqFt	21.70
.10	SANDBLASTED - 3/16"	SqFt	9.50
.11	SHEET GLASS		
	Light Sheet or Window Glass - Double Strength	SqFt	4.60
	Heavy Sheet or Crystal	SqFt	6.30
.12	STAINED GLASS - Abstract	SqFt	83.00
	Symbolism	SqFt	120.00
	Figures	SqFt	280.00
.13	STRUCTURAL GLASS - 1/4"	SqFt	12.00
.14	TEMPERED (SAFETY PLATE)		
	1/4"	SqFt	8.60
	3/8"	SqFt	11.20
	1/2"	SqFt	23.85
	5/8"	SqFt	27.75
.15	TINTED PLATE - LIGHT REFLECTIVE		
	1/4"	SqFt	9.20
	3/8"	SqFt	12.60
	1/2"	SqFt	21.50
.16	WIRED (SAFETY)		
	1/4" - Clear	SqFt	13.85
	Obscure	SqFt	10.00

The costs below are average, priced as total contractor / subcontractor costs with an overhead and fee of 10% included. Units include fasteners, tools and equipment, fringe benefits, 30% taxes and insurance on labor, and 5% General Conditions.

		UNIT	COST
0801.0	**HOLLOW METAL** (Installation of Butts and Locksets included but no hardware material)		
.1	CUSTOM FRAMES (16 ga)		
	2'6" x 6'8" x 4 3/4"	Each	250
	2'8" x 6'8" x 4 3/4"	Each	270
	3'0" x 6'8" x 4 3/4"	Each	280
	3'4" x 6'8" x 4 3/4"	Each	300
	Add for A, B or C Label	Each	32
	Add for Side Lights	Each	145
	Add for Frames over 7'0" - Height	Each	60
	Add for Frames over 6 3/4" - Width	Each	52
	CUSTOM DOORS (1 3/8" or 1 3/4") (18 ga)		
	2'6" x 6'8" x 1 3/4"	Each	300
	2'8" x 6'8" x 1 3/4"	Each	310
	3'0" x 6'8" x 1 3/4"	Each	330
	3'4" x 6'8" x 1 3/4"	Each	335
	Add for 16 ga	Each	60
	Add for B and C Label	Each	33
	Add for Vision Panels or Lights	Each	75
.2	STOCK FRAMES (16 ga)		
	2'6" x 6'8" or 7'0" x 4 3/4"	Each	180
	2'8" x 6'8" or 7'0" x 4 3/4"	Each	185
	3'0" x 6'8" or 7'0" x 4 3/4"	Each	195
	3'4" x 6'8" or 7'0" x 4 3/4"	Each	210
	Add for A, B or C Label	Each	32
	STOCK DOORS (1 3/8" or 1 3/4" - 18 ga)		
	2'6" x 6'8" or 7'0"	Each	285
	2'8" x 6'8" or 7'0"	Each	295
	3'0" x 6'8" or 7'0"	Each	315
	3'4" x 6'8" or 7'0"	Each	335
	Add for B or C Label	Each	28

		UNIT	Hollow Core	Solid Core
0802.0	**WOOD DOORS** (Installation of Butts and Locksets included but no hardware material)			
.1	FLUSH DOORS - No Label - Paint Grade			
	Birch - 7 Ply 2'4" x 6'8" x 1 3/8"	Each	180	200
	2'6" x 6'8" x 1 3/8"	Each	185	205
	2'8" x 6'8" x 1 3/8"	Each	190	215
	3'0" x 6'8" x 1 3/8"	Each	195	230
	2'4" x 6'8" x 1 3/4"	Each	190	220
	2'6" x 6'8" x 1 3/4"	Each	195	225
	2'8" x 6'8" x 1 3/4"	Each	200	245
	3'0" x 6'8" x 1 3/4"	Each	210	245
	3'4" x 6'8" x 1 3/4"	Each	215	265
	3'6" x 6'8" x 1 3/4"	Each	220	370
	Add for Stain Grade	Each		34
	Add for 5 Ply	Each		75
	Add for Jamb & Trim - Solid - Knock Down	Each		77
	Add for Jamb & Trim - Veneer - Knock Down	Each		50
	Add for Red Oak - Rotary Cut	Each		30
	Add for Red Oak - Plain Sliced	Each		40
	Deduct for Lauan	Each		23
	Add for Architectural Grade - 7 Ply	Each		75
	Add for Vinyl Overlay	Each		55
	Add for 7' - 0" Doors	Each		23
	Add for Lite Cutouts with Metal Frame	Each		84
	Add for Wood Louvres	Each		125
	Add for Transom Panels & Side Panels	Each		160

			MATERIAL			
0802.0	**WOOD DOORS, Cont'd...**					
.1	FLUSH DOORS, Cont'd...	UNIT		20-Min	45-Min	60-Min
	Label - 1 3/4" - Paint Grade					
	2'6" x 6'8" - Birch	Each		250	280	280
	2'8" x 6'8"	Each		260	290	300
	3'0" x 6'8"	Each		270	310	310
	3'4" x 6'8"	Each		290	335	335
	3'6" x 6'8"	Each		350	345	345
.2	PANEL DOORS					
	Exterior - 1 3/4"		Pine	Fir	Birch/ Oak	Lauan
	2'8" x 6'8"	Each	520	530	740	-
	3'0" x 6'8"	Each	550	560	770	-
	Interior - 1 3/8"					
	2'6" x 6'8"	Each	450	400	580	-
	2'8" x 6'8"	Each	460	410	600	-
	3'0" x 6'8"	Each	490	420	630	-
	Add for Sidelights	Each	290	420	470	
.3	LOUVERED DOORS - 1 3/8"					
	2'0" x 6'8"	Each	305	300	630	-
	2'6" x 6'8"	Each	355	350	650	-
	2'8" x 6'8"	Each	365	360	660	-
	3'0" x 6'8"	Each	375	370	680	-
.4	BI-FOLD - 2 DOOR WITH HDWE - 1 3/8"					
	2'0" x 6'8" - Flush	Each	-	-	225	190
	2'4" x 6'8"	Each	-	-	235	200
	2'6" x 6'8"	Each	-	-	245	210
	Add for Prefinished	Each				23
	See 0804.22 for 1 1/8" Closet Bi-folds					
.5	CAFE DOORS - 1 1/8" - Pair					
	2'6" x 3'8"	Each	410	-	-	-
	2'8" x 3'8"	Each	420	-	-	-
	3'0" x 3'8"	Each	440	-	-	-
.6	FRENCH DOORS - 1 3/8"					
	2'6" x 3'8"	Each	690	550	740	-
	2'8" x 3'8"	Each	700	560	770	-
	3'0" x 3'8"	Each	720	570	780	-
.7	DUTCH DOORS					
	2'6" x 3'8"	Each	880	-	-	-
	2'8" x 3'8"	Each	940	-	-	-
	3'0" x 3'8"	Each	1,000	-	-	-
.8	PREHUNG DOOR UNITS (INCL. FRAME, TRIM, HARDWARE, and SPLIT JAMBS)					
	Exterior Entrance - 1 3/4"					
	Panel - 2'8" x 6'8"	Each	480	-	-	-
	3'0" x 6'8"	Each	510	-	-	-
	3'0" x 7'0"	Each	550	-	-	-
	Add for Insulation	Each	200			
	Add for Side Lights	Each	210			
	Interior - 1 3/8"					
	Flush, H.C. - 2'6" x 6'8"	Each	-	-	345	260
	2'8" x 6'8"	Each	-	-	355	265
	3'0" x 6'8"	Each	-	-	375	275
	Add for Int. Panel Door	Each				240

			COST			
0803.0	**SPECIAL DOORS**		Oak	Birch	Lauan	Pine
.2	CLOSET BI-FOLDING - PREHUNG STOCK AND UNFINISHED	UNIT	Flush	Flush	Flush	Panel
	Wood - 2 Door - 2'0" x 6'8" x 1 3/8"	Each	270	310	230	510
	2'6" x 6'8" x 1 3/8"	Each	285	330	240	530
	3'0" x 6'8" x 1 3/8"	Each	305	380	250	600
	4 Door - 4'0" x 6'8" x 1 3/8"	Each	345	390	300	825
	5'0" x 6'8" x 1 3/8"	Each	355	410	350	920
	6'0" x 6'8" x 1 3/8"	Each	400	420	360	1,020

		UNIT	COST
	Add for Prefinished -	Each	$33.00
	Add for Jambs and Casings -	Each	$77.00
	Metal - 2 Door - 2'0" x 6'8"	Each	155
	2'6" x 6'8"	Each	160
	3'0" x 6'8"	Each	188
	4 Door - 4'0" x 6'8"	Each	215
	5'0" x 6'8"	Each	225
	6'0" x 6'8"	Each	250
	Add for Plastic Overlay	Each	38
	Add for Louvre or Decorative Type	Each	17
	Leaded Mirror (Based on 2-Panel Unit)		
	4'0" x 6'8"	Each	760
	5'0" x 6'8"	Each	810
	6'0" x 6'8"	Each	810
.3	FLEXIBLE DOORS (M) (CARPENTERS)		
	Wood Slat (Unfinished)	SqFt	1,600
	Fabric Accordion Fold	SqFt	1,500
	Vinyl Clad	SqFt	1,900
.8	PLASTIC LAMINATE FACED (M) (CARPENTERS) - 1 3/8"	SqFt	2,000
	1 3/4"	SqFt	2,000
	Add for Decorative Type	SqFt	2,100
	Add for Label Doors	SqFt	5.25
	Add for Edge Strip - Vertical and Horizontal	Each	9.50
	Add for Machining for Hardware	Each	50.00
.10	ROLLING - DOORS & GRILLES		
	Doors - 8' x 8'	Each	2,800
	10' x 10'	Each	3,000
	Grilles - 6'-8" x 3'-2"	Each	1,000
	6'-8" x 4'-2"	Each	1,100
.11	SHOWER DOORS - 28" x 66" (CARPENTERS)	Each	240
.12	SLIDING OR PATIO DOORS (M) (CARPENTERS)		
	Metal (Aluminum) - Including Glass Thresholds & Screen		
	Opening 8'0" x 6'8"	Each	1,625
	8'0" x 8'0"	Each	1,725
	Wood		
	Vinyl Clad 6'0" x 6'10"	Each	2,000
	8'0" x 6'10"	Each	2,300
	Pine - Prefinished 6'0" x 6'10"	Each	2,000
	8'0" x 6'10"	Each	2,300
	Add for Grilles	Each	330
	Add for Triple Glazing	Each	220
	Add for Screen	Each	150
.13	SOUND REDUCTION - Metal	Each	1,950
	Wood	Each	1,300
.14	TRAFFIC DOORS (CARPENTERS)		
	Electric Powered - Hollow Metal - 8' x 8' Opening	Each	5,600
	Wood - 8' x 8' Opening	Each	5,500
	Truck Impact - Metal Clad - 8' x 8' Opening	Each	3,400
	(Double Acting) - Rubber - 8' x 8' Opening	Each	2,750
	Add for Frames	Each	840
.16	VAULT DOORS - 6'6" x 2'2" with Frame - 2 hour	Each	3,400

0805.0 METAL WINDOWS (M) (CARPENTERS OR STEEL ERECTORS) (SINGLE GLAZING INCLUDED)

All Example Sizes Below: 2'4" x 4'6"

		UNIT	COST	UNIT	COST
.1	ALUMINUM WINDOWS				
	Casement and Awning	Each	345.00	SqFt	32.80
	Sliding or Horizontal	Each	275.00	SqFt	26.10
	Double and Single-Hung or Vertical Sliding	Each	305.00	SqFt	29.00
	Projected	Each	345.00	SqFt	32.80
	Add for Screens	Each	50.00	SqFt	4.76
	Add for Storms	Each	90.00	SqFt	8.50
	Add for Insulated Glass	Each	75.00	SqFt	7.10
.2	ALUMINUM SASH				
	Casement	Each	325.00	SqFt	30.90
	Sliding	Each	275.00	SqFt	26.10
	Single-Hung	Each	275.00	SqFt	26.10
	Projected	Each	230.00	SqFt	21.85
	Fixed	Each	240.00	SqFt	22.85
.3	STEEL WINDOWS				
	Double-Hung	Each	500.00	SqFt	47.50
	Projected	Each	480.00	SqFt	45.60
.4	STEEL SASH				
	Casement	Each	385.00	SqFt	36.60
	Double-Hung	Each	440.00	SqFt	41.80
	Projected	Each	330.00	SqFt	31.35
	Fixed	Each	340.00	SqFt	32.30

0806.0 WOOD WINDOWS (M) (Carpenters) - All Units Assembled, Kiln Dry Pine, Glazed, Vinyl Clad, Weatherstripped, and Top Quality. Many Other Sizes Available.

		UNIT	COST
.1	BASEMENT OR UTILITY		
	Prefinished with Screen 2'8" x 1'4"	Each	200
	2'8" x 2'0"	Each	235
	Add for Double Glazing or Storm	Each	50
.2	CASEMENT OR AWNING		
	Operating Units - Insulating Glass		
	Single 2'4" x 4'0"	Each	450
	2'4" x 5'0"	Each	500
	Double with Screen 4'0" x 4'0" (2 units)	Each	720
	4'0" x 5'0" (2 units)	Each	935
	Triple with Screen 6'0" x 4'0" (3 units)	Each	1,150
	6'0" x 5'0" (3 units)	Each	1,250
	Fixed Units - Insulating Glass		
	Single 2'4" x 4'0"	Each	430
	2'4" x 5'0"	Each	500
	Picture 4'0" x 4'6"	Each	610
	4'0" x 6'0"	Each	770
.3	DOUBLE HUNG - & Insul. Glass with Screen - 2'6" x 3'6"	Each	430
	2'6" x 4'2"	Each	440
	3'2" x 3'6"	Each	450
	3'2" x 4'2"	Each	510
	Add for Triple Glazing	Each	64
.4	GLIDER - & Insul. Glass with Screen - 4'0" x 3'6"	Each	1,050
	5'0" x 4'0"	Each	1,250
.5	PICTURE WINDOWS with CASEMENTS - & Insul. Glass with Screen		
	9'6" x 4'10"	Each	2,200
	9'6" x 5'6"	Each	2,400

		UNIT	STATIONARY	MOVABLE
0806.0	**WOOD WINDOWS Cont'd...**			
.6	CASEMENT ANGLE BAY WINDOWS-Insulating Glass with Screens			
	30° - 5'10" x 4'2" - (3 units)	Each		1,700
	7'10" x 4'2" - (4 units)	Each		2,150
	5'10" x 5'2" - (3 units)	Each		1,800
	7'10" x 5'2" - (4 units)	Each		2,100
	45° - 5'4" x 4'2" - (3 units)	Each		1,700
	7'4" x 4'2" - (4 units)	Each		1,900
	5'4" x 5'2" - (3 units)	Each		2,050
	7'4" x 5'2" - (4 units)	Each		2,250
.7	CASEMENT BOW WINDOWS - Insulating Glass			
	6'2" x 4'2" - (3 units)	Each		1,700
	8'2" x 4'2" - (4 units)	Each		2,400
	6'2" x 5'2" - (3 units)	Each		1,950
	8'2" x 5'2" - (4 units)	Each		2,500
	Add for Triple Glazing - per unit	Each		65
	Add for Bronze Glazing - per unit	Each		65
	Deduct for Primed Only - per unit	Each		16
	Add for Screens - per unit	Each		20
.8	90° CASEMENT BOX BAY WINDOWS - Insulating Glass 4'8" x 4'2"	Each		2,400
	6'8" x 4'2"	Each		3,150
	6'8" x 5'2"	Each		3,300
.9	SKY OR ROOF WINDOWS 1'10" x 3'10"	Each	660	940
	2'4" x 3'10"	Each	770	1,100
	3'8" x 3'10"	Each	910	1,300
.10	CIRCULAR TOPS & ROUNDS - 4'0"	Each	-	720
	6'0"	Each	-	1,550
0807.0	**SPECIAL WINDOWS (M) (CARPENTERS/ STEEL WORKERS)**			
.1	LIGHT-PROOF WINDOWS	SqFt		42
.2	PASS WINDOWS	SqFt		31
.3	DETENTION WINDOWS	SqFt		44
.4	VENETIAN BLIND WINDOWS (ALUMINUM)	SqFt		36
.5	SOUND-CONTROL WINDOWS	SqFt		35
0808.0	**DOOR AND WINDOW ACCESSORIES**			
.1	STORMS AND SCREENS (CARPENTERS)			
	Windows			
	Screen Only - Wood - 3' x 5'	Each		110
	Aluminum - 3' x 5'	Each		128
	Storm & Screen Combination - Aluminum	Each		143
	Doors			
	Screen Only - Wood	Each		265
	Aluminum - 3' x 6' - 8'	Each		262
	Storm & Screen Combination - Aluminum	Each		315
	Wood 1 1/8"	Each		360
.2	DETENTION SCREENS (M) (CARPENTERS)			
	Example: 4' 0" x 7' 0"	Each		740
.3	DOOR OPENING ASSEMBLIES (L&M) (GLAZIERS)			
	Floor or Overhead Electric Eye Units			
	Swing - Single 3' x 7' Door - Hydraulic	Each		4,300
	Double 6' x 7' Door - Hydraulic	Each		6,500
	Sliding - Single 3' x 7' Door - Hydraulic	Each		5,500
	Double 5' x 7' Doors- Hydraulic	Each		7,400
	Industrial Doors - 10' x 8'	Each		7,500
.4	SHUTTERS (CARPENTERS)			
	16" x 1 1/8" x 48"	Pair		125
	16" x 1 1/8" x 60"	Pair		140
	16" x 1 1/8" x 72"	Pair		145

0810.0 FINISH HARDWARE (M) (CARPENTERS)

		UNIT	COST Painted	Bronze	Chrome
.1	BUTTS				
	3" x 3"	Pair	35	37	40
	3 1/2" x 3 1/2"	Pair	37	38	42
	4" x 4"	Pair	39	40	44
	4 1/2" x 4 1/2"	Pair	42	45	53
	4" x 4" Ball Bearing	Pair	65	67	75
	4 1/2" x 4 1/2" Ball Bearing	Pair	70	75	80

		UNIT	COST
.2	CATCHES - ROLLER	Each	40
.3	CLOSURES - SURFACE MOUNTED - 3' - 0" Door	Each	200
	3' - 4"	Each	205
	3' - 8"	Each	210
	4' - 0"	Each	225
	CLOSURES - CONCEALED OVERHEAD - Interior	Each	320
	Exterior	Each	450
	Add for Fusible Link - Electric	Each	225
	CLOSURES - FLOOR HINGES - Interior	Each	770
	Exterior	Each	800
	Add for Hold Open Feature	Each	92
	Add for Double Acting Feature	Each	310
.4	DEAD BOLT LOCK - Cylinder - Outside Key	Each	200
	Cylinder - Double Key	Each	240
	Flush - Push/ Pull	Each	64
.5	EXIT DEVICES (PANIC) - Surface	Each	770
	Mortise Lock	Each	900
	Concealed	Each	2,200
	Handicap (ADA) Automatic	Each	1,600
.6	HINGES, SPRING (PAINTED) - 6" Single Acting	Each	115
	6" Double Acting	Each	145
.7	LATCHSETS - Bronze or Chrome	Each	200
	Stainless Steel	Each	250
.8	LOCKSETS - Mortise - Bronze or Chrome - H.D.	Each	260
	Mortise - Bronze or Chrome - S.D.	Each	225
	Stainless Steel	Each	365
	Cylindrical - Bronze or Chrome	Each	235
	Stainless Steel	Each	300
.9	LEVER HANDICAP - Latch Set	Each	250
	Lock Set	Each	310
.10	PLATES - Kick - 8" x 34" - Aluminum	Each	92
	Bronze	Each	81
	Push - 6" x 15" - Aluminum	Each	52
	Bronze	Each	77
	Push & Pull Combination - Aluminum	Each	93
	Bronze	Each	135
.11	STOPS AND HOLDERS		
	Holder - Magnetic (No Electric)	Each	215
	Bumper	Each	52
	Overhead - Bronze, Chrome or Aluminum	Each	110
	Wall Stops	Each	35
	Floor Stops	Each	46
	See 0808.3 for Automatic Openers & Operators		

0811.0 WEATHERSTRIPPING (L&M) (CARPENTERS)

		UNIT	COST
.1	ASTRAGALS - Aluminum - 1/8" x 2"	Each	46
	Painted Steel	Each	46
.2	DOORS (WOOD) - Interlocking	Each	86
	Spring Bronze	Each	98
	Add for Metal Doors	Each	46
.3	SWEEPS - 36" WOOD DOORS - Aluminum	Each	40
	Vinyl	Each	30
.4	THRESHOLDS - Aluminum - 4" x 1/2"	Each	46
	Bronze 4" x 1/2"	Each	70
	5 1/2" x 1/2"	Each	76
.5	WINDOWS (WOOD) - Interlocking	Each	78
	Spring Bronze	Each	98

All items in this division are subcontractors' costs and include all labor, materials, equipment and fees.

0901.0 LATH & PLASTER (L&M) (Latherers and Plaster)

		UNIT	COST
.1	LATHING		
.11	Metal Lath	SqYd	15.50
.12	Gypsum Lath	SqYd	13.25
.13	Metal Studs	SqYd	17.00
.14	Channels - Studs - 16" O.C.	SqYd	16.50
	Furring - Beams & Columns - 16" O.C.	SqYd	20.00
	Ceilings (including hangers) - 16" O.C.	SqYd	23.00
.15	Metal Trim - Base Bead and Corner Bead	LnFt	3.80
	Picture Mould	LnFt	5.10
.2	PLASTERING		
.21	Gypsum (Putty Coat, Sand Float Textured),Walls - 1 Coat	SqYd	17.00
	3 Coat	SqYd	48.00
	Add for Ceiling, Column and Beam Applications	SqYd	4.50
.22	Acoustical	SqYd	28.00
.23	Vermiculite	SqYd	25.00
.24	Stucco or Cement Plaster		
	On Masonry - Large Areas - Float Finish - 1 Coat	SqYd	28.50
	3 Coat	SqYd	37.00
	Panels, Facias, Soffits - 3 Coat	SqYd	44.00
	On Metal Lath - Large Areas - 3 Coat	SqYd	49.00
	Panels, Facias, Soffits, Columns	SqYd	59.00
	On Metal Studs - 4" - 16 ga	SqYd	35.00
	6" - 16 ga	SqYd	50.00
	Deduct for 18 ga Studs	SqYd	10%
	Add for Trowel Finish	SqYd	4.60
.25	Fireproofing (Sprayed On) - Mineral Fiber		
	Steel and Decks - 1 Hour	SqYd	10.50
	2 Hour	SqYd	17.70
	Beams and Columns - Not Wrapped - 2 Hour	LnFt	18.60
	3 Hour	LnFt	21.50
.26	Thin Coat or Veneer Plaster - Walls - 1 Coat	SqYd	15.00
	2 Coat	SqYd	23.00
	Ceilings - 2 Coat	SqYd	29.00
	Add for Patching	SqYd	50%
	Add Color to Plaster and Stucco	SqYd	5.25
	Add for Scaffold - Average	SqYd	9.20
.27	Exterior Wall System - Stucco or Synthetic Plaster		
	Includes Structural Board and 2" Insulation	SqYd	107.00
	Add for Esthetic Grooving	LnFt	5.00

LATH AND GYPSUM PLASTER IN COMBINATION UNITS:		FINISHED	
WALLS	UNIT	1 Side	2 Side
1 Coat Plaster on Waterproofed Found. Walls	SqYd	17.20	-
2 Coat Plaster on Masonry and Concrete	SqYd	34.00	68.00
on Gypsum Lath	SqYd	42.50	97.00
3 Coat Plaster on Metal Lath	SqYd	53.00	108.00
2 Coat Plaster on Metal Studs & Gypsum Lath	SqYd	70.00	115.00
CEILINGS			
3 Coat Plaster on Metal Lath and Channel for Suspended Ceiling (including hangers)	SqYd	100.00	-
3 Coat Plaster and Metal Lath attached to joists, beams, etc.	SqYd	84.00	-
FURRED			
Beam - 3 Coat Plaster, Metal Lath, Channel	SqYd	102.00	-
Column & Pipe - 3 Coat Plaster - Metal Lath	SqYd	96.00	-
Cabinet - 3 Coat Plaster, Metal Lath, Chnl.	SqYd	65.00	-
2 Coat Plaster on Gypsum Lath	SqYd	59.00	-

0902.0 GYPSUM DRYWALL (L&M) (Carpenters)
Based on 8' Ceiling Heights & Screwed-on Application

		UNIT	COST
.1	BACKER BOARD		
	3/8"	SqFt	1.05
	1/2"	SqFt	1.08
	5/8" Fire Rated	SqFt	1.09
	Deduct for Nailed-on	SqFt	.08
.2	FINISH BOARD - Walls		
	1/2" - Standard	SqFt	1.65
	Fire Rated	SqFt	1.68
	Moisture Resistant	SqFt	1.76
	Insulating (Foil Backed)	SqFt	1.81
	Combined Fire Rated and Insulating	SqFt	1.85
	5/8" - Standard	SqFt	1.65
	Fire Rated	SqFt	1.67
	Moisture Resistant	SqFt	1.67
	Insulating (Foil Backed)	SqFt	1.70
	Combined Fire Rated and Insulating	SqFt	1.80
	Lead Lined with #4 Lead - 1/16"	SqFt	7.75
	#2 Lead - 1/16"	SqFt	6.10
	Add for Ceiling Work - to Wood	SqFt	.17
	Add for Ceiling Work - to Steel	SqFt	.40
	Add for Adhesive Method	SqFt	.17
	Deduct for Nailed On	SqFt	.08
	Add for Beam and Column Work	SqFt	.75
	Add for Resilient Clip Application	SqFt	.85
	Add for Small Cut Up Areas	SqFt	.62
	Add for Each Floor Added	SqFt	.07
	Add for Work Above 10'	SqFt	.25
	Add for Filling Hollow Metal Frames	Each	75.00
.3	VINYL COATED BOARD - 1/2"	SqFt	2.05
	5/8"	SqFt	2.15
.4	LINER PANELS - 1"	SqFt	2.75

			COST	
.5	STEEL STUDS (Incl. Top & Bottom Plates)	UNIT	25 ga	20 ga
	16" O.C. - 1 5/8"	SqFt	1.40	1.61
	2 1/2"	SqFt	1.42	1.63
	3 5/8"	SqFt	1.38	1.58
	6"	SqFt	1.44	1.64
	24" O.C. - 1 5/8"	SqFt	1.08	1.31
	2 1/2"	SqFt	1.12	1.34
	3 5/8"	SqFt	1.13	1.39
	6"	SqFt	1.20	1.49
	Add for Work Over 10' Heights	SqFt	.17	.17
	Add for 16 Ga 6" Galvanized	SqFt	1.65	1.23
.6	FURRING CHANNELS			
	16" O.C. 3/4" x 1 3/8" walls	SqFt	-	1.18
	24" O.C. 3/4" x 1 3/8" walls	SqFt	-	1.08
	Resilient Channels - 16" O.C.	SqFt	-	1.08
	Add for Ceiling Work	SqFt	-	.15
	Add for Beam and Column Work	SqFt	-	1.13
.7	METAL TRIM - Casing Bead	LnFt	-	1.13
	Corner Bead	LnFt	-	.88
.8	TAPING AND SANDING	SqFt	-	.52
.9	TEXTURING	SqFt	-	.60

See 0602.34 for Labor & Material Priced Separately

0902.0 GYPSUM DRYWALL, Cont'd...

COMBINATION UNITS - METAL FRAMING & BOARD

	UNIT	COST
PARTITIONS 3 5/8" - 25 ga Studs - 24" O.C.		
1/2" Board One Side	SqFt	2.90
Two Sides	SqFt	4.05
1/2" Fire Rated Board Two Sides	SqFt	4.10
1/2" Fire Rated Board One Side - 2 Layers 2nd Side (1 Hr.)	SqFt	5.10
2 Layers Each Side (2 Hr.)	SqFt	6.00
5/8" Board One Side	SqFt	2.93
Two Sides	SqFt	4.10
5/8" Fire Rated Board Each Side (1 Hr.)	SqFt	4.15
5/8" Fire Rated Board - 2 Layers Each Side (2 Hr.)	SqFt	6.30
3 Layers to Columns & Beams (2 Hr.)	SqFt	9.00
Add for Studs 16" O.C. - 25 Ga	SqFt	.27
Deduct for 1 5/8" Studs - 25 Ga	SqFt	.24
Deduct for 2 5/8" Studs - 25 Ga	SqFt	.13
Add for Studs - 20 Ga and 16" O.C.	SqFt	.30
Add for Studs - 20 Ga and 24" O.C.	SqFt	.28
Add for 6" Studs - 25 Ga and 16" O.C.	SqFt	.13
Add for 6" Studs - 20 Ga and 16" O.C.	SqFt	.22
Add for Sound Deadening	SqFt	.85
PARTITIONS - CAVITY SHAFT WALL -		
25 Ga Studs - 24" O.C. - All Board Fire Code C		
1" Liner Board One Side - 5/8" Board 2nd Side (1 Hr.)	SqFt	5.35
2 Layers - 1/2" Board 2nd Side (2 Hr.)	SqFt	5.85
3 Layers - 5/8" Board 2nd Side (3 Hr.)	SqFt	7.00
1" Liner Board and 5/8" Board One Side		
2 Layers 1" Liner Board 2nd Side	SqFt	9.40
PARTITIONS - DEMOUNTABLE		
2 1/2" Studs 24" O.C.-1/2" Vinyl Coated Bd. 2 Sides	SqFt	4.90
3 5/8" Studs 24" O.C.-1/2" Vinyl Coated Bd. 2 Sides	SqFt	5.30
Add for 16" O.C.	SqFt	.33
Add for 5/8" Vinyl Coated Board 2 Sides	SqFt	.36
Add for Special Colors Baked On	SqFt	25%
FURRED WALLS		
16" O.C. 1/2" Board	SqFt	2.68
24" O.C. 1/2" Board	SqFt	2.75
Resilient Channels and 1/2" Board	SqFt	2.70
CEILINGS		
To Wood Joists - 1/2" Fire Rated Board (1 Hr.)	SqFt	1.35
To Steel Joists with Furring Channels		
24" O.C. - 1/2" F.R. Board (2 Hr. with 2 1/2" Conc.)	SqFt	2.60
5/8" F.R. Board (1 Hr. with 2" Conc.)	SqFt	2.65
12" O.C. - 5/8" F.R. Board (2 Hr. with 2 1/2" Conc.)	SqFt	3.20
COLUMNS - 14 WF and Up		
1/2" F.R. Board with Screw Studs at Corners (1 Hr.)	SqFt	3.35
2 Layers - 1/2" F.R. Bd. with Screw Studs @ Corners (2 Hr.)	SqFt	3.80
Add for Corner Bead	LnFt	1.20

See 0602.34 for Labor and Material Priced Separately
See 0703.0 for Insulation Work

		UNIT	COST
0903.0	**TILEWORK (L&M) (Tile Setters)**		
.1	CERAMIC TILE		
	Floors - 1" Square - Thin Set (Unglazed)	SqFt	8.40
	- 1" Hex - Thin Set	SqFt	9.10
	1" x 2"	SqFt	9.20
	2" Square	SqFt	8.60
	16" Square	SqFt	9.20
	Add for Mortar Setting Bed	SqFt	4.90
	Add for Random Colors	SqFt	1.16
	Add for Epoxy Joints	SqFt	2.50
	Conductive - In Mortar Setting Bed	SqFt	13.00
	Walls - 2" x 2" - Thin Set	SqFt	10.30
	4 1/4" x 4 1/4"	SqFt	7.85
	4 1/4" x 6"	SqFt	7.95
	4 1/4" x 8"	SqFt	8.25
	6" x 6"	SqFt	8.70
	Add for Plaster Setting Bed	SqFt	4.75
	Add for Decorator Type	SqFt	6.80
	Base - 4 1/4"	LnFt	8.10
	6"	LnFt	8.15
	Add for Coved Base	LnFt	.97
	Accessories - Toilet and Bath Embedded	Each	60.00
.2	QUARRY TILE		
	Floors - 4" x 4" - Thin Set	SqFt	8.50
	6" x 6"	SqFt	8.40
	4" x 8"	SqFt	8.40
	Hexagon, Valencas, etc.	SqFt	9.60
	Walls - 4" x 4" - Thin Set	SqFt	9.20
	6" x 6"	SqFt	9.00
	Stairs	SqFt	12.00
	Base - 4" x 4" x 1/2"	LnFt	8.80
	6" x 6" x 1/2"	LnFt	9.80
	Add for Hydroment Joints	SqFt	.90
	Add for Epoxy Joints	SqFt	2.40
	Add for Non-Slip Tile	SqFt	.54
	Add for Mortar Setting Bed	SqFt	4.80
0904.0	**TERRAZZO (L&M) (Terrazzo Workers) - Based on Gray Cement, 4' Panel, and White Dividing Strips**		
.1	CEMENT TERRAZZO		
	Floors - 2" Bonded	SqFt	12.50
	2 1/2" Not Bonded	SqFt	11.00
	Base - 6" without Grounds or Base Bead	LnFt	19.70
	6" with Grounds and Base Bead	LnFt	25.00
	Stairs - Treads and Risers - Tread Length	LnFt	51.00
	Stringers without Base Bead	LnFt	45.00
	Pan Filled Treads and Landings	LnFt	21.00
	Add for White Cement	SqFt	.75
	Add for Brass Dividing Strips	SqFt	1.10
	Add for Large Chips (Venetian) - 3"	SqFt	6.00
	Add for 2-Color Panels	SqFt	3.60
	Add for 3-Color Panels	SqFt	4.25
	Add for Non-Slip Abrasives	SqFt	1.00
	Add for Small or Scattered Areas	SqFt	15%
.2	CONDUCTIVE TERRAZZO - 2" Bonded	SqFt	14.00
	1/4" Thin Set Epoxy	SqFt	12.20
.3	EPOXY TERRAZZO - 1/4" Thin Set	SqFt	10.10
.4	RUSTIC TERRAZZO (WASHED AGGREGATE) - Marble	SqFt	7.00
	Quartz	SqFt	7.60

See 0907 for Precast Terrazzo Tiles

0905.0 ACOUSTICAL TREATMENT (L&M) (Carpenters)

		UNIT	COST
.1	MINERAL TILE (No Backing or Supports)		
	Fissured or Perforated - 1/2" x 12" x 12" or 12" x 24"	SqFt	2.20
	5/8" x 12" x 12" or 12" x 24"	SqFt	2.50
	3/4" x 12" x 12" or 12" x 24"	SqFt	2.60
	Add to all above for Backing or Supports if needed:		
	Ceramic Faces	SqFt	.52
	Concealed Z Spline Application	SqFt	.66
	Wood Furring Strips	SqFt	.73
	Rock Lath	SqFt	.95
	1 1/2" Channel Suspension	SqFt	1.35
	Resilient Furring Channels	SqFt	.95
.2	PLASTIC COVERED TILE (No Backing or Supports)		
	3/4" x 12" x 12"	SqFt	2.45
	5/8" x 12" x 12"	SqFt	2.05
.3	METAL PAN TILE (Supports Included)		
	12" x 24" or 12" x 36" or 12" x 48" - Steel (24 Ga)	SqFt	7.50
	Aluminum (.025)	SqFt	9.25
	Aluminum (.032)	SqFt	10.50
	Stainless Steel	SqFt	12.60
	Add for Color	SqFt	.54
.4	CLASS A MINERAL TILE		
	(Grid System with Suspension incl. & Partitions to Clg.)		
	5/8" x 24" x 24" or 24" x 48" - Perforated	SqFt	1.77
	Fissured	SqFt	1.78
	Add for Partitions thru Ceilings	SqFt	.27
	Add for Translucent Panels for Lights	SqFt	1.50
.5	FIRE RATED DESIGN MINERAL TILE		
	(Grid System with Suspension incl. & Partitions to Clg.)		
	5/8" x 24" x 24" or 24" x 48" - Perforated (2-hour rating)	SqFt	1.93
	Fissured (2-hour rating)	SqFt	1.94
	3/4" x 24" x 24" or 24" x 48" - Perforated (3-4 Hr rating)	SqFt	2.10
	Fissured (3-4 Hr rating)	SqFt	2.11
	Add for Partitions thru Ceilings	SqFt	.27
	Add for Reveal Edge or Shadow Line (24" x 24" only)	SqFt	.37
	Add for Nubby Tile	SqFt	1.35
	Add for Ceramic Faces (5/8")	SqFt	.81
	Add for Colored Tile	SqFt	.27
	Add for Fixture Fire Protection - Fixtures only	SqFt	1.12
.6	GLASS FIBRE BOARD (Grid System with Suspension included)		
	5/8" x 24" x 24"	SqFt	1.90
	3/4" x 24" x 48"	SqFt	1.95
	1" x 48" x 48"	SqFt	1.90
	1-1/2" x 60" x 60"	SqFt	3.35
	Add for Three-Dimensional	SqFt	3.25
.7	SOUND CONTROL PANELS AND BAFFLES	SqFt	13.20

		UNIT	COST
.8	ACOUSTICAL WALL PANELS - Special		
	Fiberglass with Cloth Face - 3/4"	SqFt	4.30 to 10.80
	1"	SqFt	5.50 to 12.40
	1-1/2"	SqFt	8.50 to 14.50
	3/4" Mineral	SqFt	1.40 to 3.85
	1/2" Fibre Board	SqFt	.90 to 1.35
.9	LINEAR METAL CEILINGS		
	4" Wide - Painted	SqFt	6.20 to 8.15
	Chrome or Brass	SqFt	9.00 to 11.40

See 0307 for Fibre Cementitious Acoustical Plank

0906.0 VENEER STONE (Floors, Walls, Stools, Stairs, Partitions) (2" and Under) (L&M) (Marble Setters)

		UNIT	COST
.1	LIMESTONE		
.11	Cut Stone - Floors 7/8" Interior	SqFt	31.50
	Floors 2" Exterior	SqFt	35.70
	Walls 1"	SqFt	34.60
	Stools 1" x 8"	LnFt	41.30
.12	Flagstone Floors 1"	SqFt	38.00
.2	MARBLE Floors 5/8"	SqFt	32.80
	Floor Tiles 3/8" x 12" x 12"	SqFt	34.80
	Walls 3/8" x 12" x 12"	SqFt	38.30
	Base 1" x 6"	LnFt	28.30
	Stool 1" x 8"	LnFt	39.00
	Stairs - Treaded Risers	SqFt	75.50
.3	SLATE Floors 3/8" Package	SqFt	21.00
	Floors 3/4" Random or Flagstone	SqFt	22.60
	Walls 3/4" Random	SqFt	28.10
	Stools 1" x 8"	LnFt	40.00
.4	SANDSTONE Floors 2"	SqFt	35.80
.5	GRANITE Floors 5/6" Interior	SqFt	41.20
	Walls - 2" Exterior	SqFt	50.50
	1 1/2" Interior	SqFt	46.30
	Counter Tops 1 1/4"	SqFt	69.20
.6	SYNTHETIC MARBLE Floors 5/8"	SqFt	18.90
	Walls 5/8"	SqFt	21.60
	Add for Leveling Old Floors, per Square Foot		2.50

0907.0 SPECIAL FLOORING (L&M)

		UNIT	COST
.1	MAGNESIUM OXYCHLORIDE (Seamless)	SqFt	9.10
.2	ELASTOMERIC	SqFt	10.40
.3	SYNTHETIC		
	Epoxy with Granules 1/16" Single Speed	SqFt	5.20
	3/32" Double Speed	SqFt	6.30
	Epoxy with Aggregates 1/4" Industrial	SqFt	8.25
	Polyester with Granules 3/32"	SqFt	6.80
	1/4"	SqFt	8.50
	See 0904.3 for Epoxy Terrazzo and 0603.2 for Counter Tops		
.4	ASPHALTIC MASTIC	SqFt	8.30
	Add for Acid Proof	SqFt	1.50
.5	BRICK PAVERS - Mortar Setting Bed Application		
	3 7/8" x 8" x 1 1/4" - Red	SqFt	10.00
	Iron Spot	SqFt	11.40
	Acid Proof	SqFt	21.70
	See 0206.3 for Patios and Walks on Sand Cushion		
.6	PRECAST PANELS		
	12" x 12" x 1" Precast Terrazzo	SqFt	27.00
	12" x 12" x 1" Precast Exposed Aggregate	SqFt	27.20
	12" x 12" x 1" Ceramic (Terra Cotta)	SqFt	32.50
	6" Precast Terrazzo Base	LnFt	21.30
.7	NEOPRENE 1/4"	SqFt	12.00

0908.0 RESILIENT FLOORING (L&M)
(Carpet and Resilient Floor Layers)

		UNIT	COST
.1	ASPHALT - Tile 9" x 9" x 1/8"	SqFt	3.00
.2	CORK - Tile 1/8" - Many Sizes	SqFt	4.50
	3/16" - Many Sizes	SqFt	4.75
.3	RUBBER		
	Tile 12" x 12" x 3/32"	SqFt	5.50
	12" x 12" x 3/32" Conductive	SqFt	6.50
	Base 2 1/2"	LnFt	1.60
	4"	LnFt	1.85
	6"	LnFt	2.00
	Treads, Stringers and Risers		
	Treads 3/16" Gauge Pre-formed	LnFt	13.00
	1/4" Gauge Pre-formed	LnFt	15.00
	5/16" Gauge Pre-formed	LnFt	19.00
	Stringers 10" x 1/8"	LnFt	10.00
	Risers	LnFt	5.50
.4	VINYL COMPOSITION TILE 12" x 12" x 3/32"	SqFt	1.60
	12" x 12" x 1/8"	SqFt	1.75
	Conductive	SqFt	6.00
	Treads 1/8"	LnFt	6.40
	3/16"	LnFt	8.00
	1/4"	LnFt	10.00
	Stringers 10" x 1/8"	LnFt	10.00
	Risers 7" x 1/8"	LnFt	4.50
	Add for Marbleized or Patterns	SqFt	1.40
.5	VINYL TILE 9" x 9" x 1/8" Solid	SqFt	6.80
	12" x 12" x 1/8"	SqFt	7.90
	Base 2 1/2"	LnFt	1.63
	4"	LnFt	1.68
	6"	LnFt	1.90
.6	VINYL SHEET 1/8" Inlaid	SqFt	2.80 – 7.30
	1/8" Homogenous	SqFt	4.50 – 8.00
	Acoustic Cushioned	SqFt	4.90 – 11.40
	Coving	LnFt	6.50
.7	VINYL CORNER GUARDS	LnFt	5.80
	Add Plywood or Masonite Underlayment	SqFt	1.72
	Add Diagonal Cuttings	SqFt	.60
	Add Clean and Wax	SqFt	.45
	Add Skim Coat Leveling	SqFt	2.00 – 3.70

0909.0 CARPETING (L&M) (Carpet Layers)

		UNIT	CUT PILE	LOOP PILE
.1	NYLON - 20 oz	SqYd		14.60 - 25.00
	22 oz	SqYd		15.60 - 26.00
	24 oz	SqYd	13.00 – 19.00	16.70 - 30.00
	26 oz	SqYd		17.80 - 30.00
	28 oz	SqYd		18.00 - 32.00
	30 oz	SqYd	15.60 – 23.00	19.60 - 35.00
	32 oz	SqYd	16.80 – 24.00	21.00 - 37.00
.2	WOOL - 36 oz	SqYd		27.30 - 68.00
	42 oz	SqYd		34.60 - 80.00
.3	OLEFIN - 20 oz	SqYd		13.00 - 22.00
	22 oz	SqYd		14.20 - 23.00
	24 oz	SqYd		15.70 - 25.00
	26 oz	SqYd		16.80 - 28.00
	28 oz	SqYd		18.40 - 30.00
.4	CARPET TILE, MODULAR-Tension Bonded	SqYd		36.50 - 69.00
	Tufted	SqYd	25.00 – 38.60	25.00 - 50.00
.5	INDOOR-OUTDOOR	SqYd		10.00 - 23.00

Installation included in above prices

		UNIT	COST
0909.0	**CARPETING, Cont'd...**		
	Add to Section 0909.0 above:		
	Add for Custom Carpet	SqYd	10.00 - 23.00
	Add for Cushion (Urethane, Rubber, Synthetic)	SqYd	4.20 - 7.00
	Add Bonded Cushion (Urethane Backed)	SqYd	1.45 - 1.90
	Add for Carpet Base, Top Edge Bound	LnFt	2.20 - 3.50
	Add for Skim Coat Leveling, Latex	SqFt	1.20 - 1.75
	Add for Border and Inset Work	LnFt	2.60 - 4.50
	Add for Reduction Strips	LnFt	1.90 - 3.00

		UNIT	COST
0910.0	**WOOD FLOORING (L&M) (Carpenters)**		
.1	STRIP AND PLANK FLOORING		
	Oak Floors (based on Standard Grade)		
	White Oak 25/32 1 1/2" Select Job Finished	SqFt	10.50
	2 1/4" Select Job Finished	SqFt	10.70
	3 1/4" Select Job Finished	SqFt	11.00
	Add for Clear Grade	SqFt	.40
	Deduct for Antique or Rustic Floors	SqFt	.50
	Maple Floors - Iron Bound (Edge Grain)		
	3/4" x 1 1/2" #2 & Better Gr Job Finished	SqFt	11.20
	3/4" x 2 1/4" #2 & Better Gr Job Finished	SqFt	11.80
	33/32 x 2 1/4" #2 & Better Gr Job Finished	SqFt	11.20
	33/32 x 2 1/4" #3 Grade Job Finished	SqFt	10.50
	Add for #1 Grade	SqFt	1.20
	Add to above for Cork Underlayment	SqFt	1.30
	Fir Floors		
	1" x 4" Vertical Grain B&B Job Finished	SqFt	5.50
	Flat Grain B&B Job Finished	SqFt	5.30
.2	WOOD PARQUET Pre Finished 5/16" Oak or Maple	SqFt	8.60
	FLOORING Pre Finished 3/4 " Oak or Maple	SqFt	10.60
	Pre Finished 5/16" Teak	SqFt	11.50
	5/16" Walnut	SqFt	11.80
.3	RADIANT TREATED (Plastic and Wood) 5/16" x 12" x 12"	SqFt	12.90
.4	RESILIENT WOOD FLOOR SYSTEMS		
	Cushion Sleeper Floors (Maple)		
	33/32 2 1/4" #1 Grade Pre-Finished	SqFt	9.50
	#2 Grade Pre-Finished	SqFt	9.10
	#3 Grade Pre-Finished	SqFt	8.40
	Cork Underlayment included	SqFt	1.35
	Add for Plywood Underlayment (2 layers)	SqFt	2.10
	Fixed Sleeper Floors (Oak)		
	33/32 2 1/4" #1 Grade Pre-Finished	SqFt	9.50
	#2 Grade Pre-Finished	SqFt	9.35
	1/4" #3 Grade Pre-Finished	SqFt	9.25
.5	WOOD BLOCK INDUSTRIAL FLOORS (Creosoted) 2"	SqFt	5.00
	2 1/2"	SqFt	5.80
	3"	SqFt	6.00
.6	FINISHING (Sanding, Sealing and Waxing)		
	Included in Job Finished Prices Above for New House	SqFt	1.95
	for New Gym	SqFt	1.90
	for Old Gym	SqFt	3.20
	Add to Section 0910.0 above:		
	Small Cut-up Areas	SqFt	30%
	Trim and Irregular Areas	SqFt	15%
	Polyurethane Finish with Stain	SqFt	20%
	Deduct for Residential Work	SqFt	10%

		UNIT	COST
0911.0	**SPECIAL COATINGS**		
.1	CEMENTITIOUS (Glazed Cement)	SqFt	2.70 - 4.40
.2	EPOXIES	SqFt	1.60 - 3.05
.3	POLYESTERS	SqFt	1.65 - 3.15
.4	VINYL CHLORIDE	SqFt	1.90 - 3.45
.5	AGGREGATE TO EPOXY - Blown on Quartz	SqFt	8.00
	Troweled on Quartz	SqFt	8.30

		UNIT	COST
.6	SPRAYED FIREPROOFING (with Perlite)		
	Column - 1" 2 Hour	LnFt	1.80
	1 3/8" 3 Hour	LnFt	2.10
	1 3/4" 4 Hour	LnFt	2.65
	Beams - 1 1/8" 2 Hour	LnFt	2.00
	1 1/4" 3 Hour	LnFt	2.10
	1 1/2" 4 Hour	LnFt	2.35
	Ceilings - 1/2" 1 Hour	SqFt	1.50
	5/8" 2 Hour	SqFt	1.55
	1" 3 Hour	SqFt	1.90

See 0901.25 for Other Fireproofing

		UNIT	COST
0912.0	**PAINTING (L&M) (Painters)**		
	WALLS AND CEILINGS - BRUSH		
	Brick 2 Coats	SqFt	.75
	Concrete 2 Coats	SqFt	.68
	Concrete Block - Interior 2 Coats	SqFt	.71
	Exterior 2 Coats	SqFt	.73
	Light Weight 2 Coats	SqFt	.76
	2 Coats/Filler	SqFt	.93
	Plaster 2 Coats	SqFt	.70
	Sheet Rock 2 Coats	SqFt	.60
	Steel Decking & Sidings 1 Coat	SqFt	.51
	Stucco 2 Coats	SqFt	.86
	Wood Siding & Shingles 2 Coats	SqFt	.84
	Wood Paneling - Stain, Seal and Varnish	SqFt	1.55
	Deduct for Roller Application	SqFt	.11
	Deduct for Spray Application	SqFt	.16
	Add for Scaffold Work	SqFt	.25
	Add for Epoxy Paint	SqFt	.23
	DOORS AND WINDOWS		
	Door, Wood - 2 Coats	Each	44.00
	Stain, Seal and Varnish	Each	65.00
	Bi-fold Door - Stain, Seal and Varnish 48"	Each	66.00
	60"	Each	73.00
	72"	Each	87.00
	Door Frame, Wood (incl. Trim) - 2 Coats	Each	40.00
	Stain, Seal & Varnish	Each	50.00
	Door, Metal (Factory Primed) - 1 Coat	Each	36.00
	2 Coats	Each	48.00
	Door Frame, Metal (Factory Primed) - 1 Coat	Each	33.00
	2 Coats	Each	37.00
	Window, Wood (including Frame & Trim) - 2 Coats	Each	49.00
	Add for Storm Windows or Screens	Each	26.00
	Add for Shutters - Pair	Pair	70.00
	Window, Steel	Each	44.00
	FLOORS, EPOXY	SqFt	2.10

0912.0 PAINTING (L&M) (Painters)	UNIT	COST
MISCELLANEOUS WOOD		
Cabinets - Enamel 2 Coats	SqFt	.90
Stain and Varnish	SqFt	1.10
Enamel 3 Coats	SqFt	1.35
Stain, Seal and Varnish	SqFt	1.70
Wood Trim - Stain, Seal and Varnish - Large	LnFt	1.70
Small or Decorative	LnFt	2.00
MISCELLANEOUS METAL - 1 Coat Sprayed		
Miscellaneous Iron and Small Steel	Ton	1.10
Structural Steel	Ton	.74
Structural Steel - 2 Sq.Ft. per Ln.Ft.	LnFt	.56
3 Sq.Ft. per Ln.Ft.	LnFt	.69
4 Sq.Ft. per Ln.Ft.	LnFt	.95
5 Sq.Ft. per Ln.Ft.	LnFt	1.15
Add for Brush	LnFt	40%
Add for 2 Coats	LnFt	75%
Gutters, Down Spouts and Flashings - 4"	LnFt	1.50
5"	LnFt	1.55
6"	LnFt	1.75
Piping - Prime and 1 Coat - 4"	LnFt	.74
6"	LnFt	1.05
8"	LnFt	1.20
10"	LnFt	1.65
Radiators	SqFt	3.15
MISCELLANEOUS		
Gold and Silver Leaf	SqFt	46.00
Parking Lines - Striping	LnFt	.33
Wood Preservatives	SqFt	.56
Waterproof Paints - 2 Coats	SqFt	.96
Clear - 1 Coat	SqFt	.50
2 Coat	SqFt	.75
Taping and Sanding	SqFt	.50
Texturing	SqFt	.55

See 0910 for Wood Floor Finishing

0913.0	WALL COVERINGS (L&M) (Painters)	UNIT	COST
.1	VINYL PLASTICS (54" Roll)		
	Light Weight	SqFt	1.54 - 2.00
	Medium Weight	SqFt	1.60 - 2.50
	Heavy Weight	SqFt	1.70 - 4.50
	Add for Sizing	SqFt	.20 - .38
	Add for 27" and 36" Rolls	SqFt	.17 - .40
.2	PLASTIC WALL TILE	SqFt	3.10 - 4.50
.3	WALLPAPER ($10.00-40.00 Roll, 36 S.F. to Roll)	SqFt	1.50 - 4.00
	Add or Deduct per Roll - $1.00	SqFt	.20 - .79
.4	FABRICS (Canvas, etc.)	SqFt	1.60 - 3.50
.5	LEATHER	SqFt	2.15 - 3.65
.6	WOOD VENEERS	SqFt	5.50 - 10.50
.7	FIBERGLASS	SqFt	2.25 - 4.00
.8	CORK - 3/16"	SqFt	1.87 - 4.50
.9	GLASS - 5/16"	SqFt	1.25 - 2.40
.10	GRASS CLOTH	SqFt	1.80 - 3.25

				MATERIAL		
1001.0	**ACCESS PANELS (M) (Lathers, Carpenters)**	UNIT	LABOR	Flush Flange	Drywall	UL Label
	10" x 10"	Each	16.00	36.00	40.00	117.00
	12" x 12"	Each	16.70	42.00	-	122.00
	14" x 14"	Each	17.70	45.00	50.00	145.00
	16" x 16"	Each	19.30	47.00	-	163.00
	18" x 18"	Each	21.80	56.00	-	167.00
	22" x 22"	Each	24.00	73.00	84.00	200.00
	24" x 24"	Each	28.00	86.00	-	220.00
	24" x 36"	Each	32.50	100.00	-	280.00
	32" x 32"	Each	36.50	117.00	-	300.00
	Add for Plaster Type					15%
	Add for Locks					9.00

		UNIT	LABOR	MATERIAL
1002.0	**CHALKBOARDS AND TACKBOARDS (Carpenters)**			
	Chalkboards			
	Slate - Natural 3/8" - New	SqFt	4.20	17.50
	Resurfaced	SqFt	4.20	9.00
	Porcelain Enameled Steel			
	1/2" Foil Backed Gypsum and 24 Ga	SqFt	3.70	8.50
	28 Ga	SqFt	3.70	8.00
	Marker (White) Board	SqFt	3.70	12.00
	Tackboard			
	Cork - 1/4" on 1/4" Gypsum Backing - Washable	SqFt	1.82	7.00
	Add for Colored Cork	SqFt	-	3.10
	Deduct for Natural Style - Not Washable	SqFt	-	3.00
	Trim (Aluminum)			
	Trim-Slip & Snap (incl. Metal Grounds) 1/4#	LnFt	2.62	2.10
	1/2#	LnFt	2.95	4.25
	Map Rail - 3/4 lb/ft	LnFt	3.25	5.80
	1/4 lb/ft	LnFt	2.95	3.30
	Chalk Tray	LnFt	4.10	7.00
	Average Trim Cost	SqFt	3.10	6.30
	Combination Unit Costs (Average)			
	Slate - 3/8" - with Tackboard and Trim	SqFt	4.20	15.00
	Enameled Steel (24 Ga) with Tackbd. & Trim, 8' x 4'	SqFt	4.00	9.00
	Marker (White) Board	SqFt	4.10	9.50
	Tackboard with Trim	SqFt	4.20	12.00
	Add Vertical and Horizontal Sliding Units	SqFt	-	5.80
	Portable Chalkboard Tackboard - 4' x 6'	Each	67.00	950.00
	Add for Revolving Type		-	320.00
1003.0	**CHUTES (Laundry & Rubbish)(M)(Sheet Metal Workers)**			
	Laundry or Linen			
	(16 Ga Aluminized Steel and 10' Floor Height)			
	1 Floor Opening - 24" Dia - per Opening	Opng	230.00	1,000.00
	2 Floors or Openings	Opng	215.00	825.00
	6 Floors or Openings	Opng	205.00	775.00
	Average	LnFt	22.00	87.00
	Rubbish			
	(16 Ga Aluminized Steel)			
	1 Floor Opening - 24" Dia - per Opening	Opng	225.00	1,000.00
	2 Floors or Openings	Opng	215.00	860.00
	6 Floors or Openings	Opng	205.00	750.00
	Add for Explosion Vent - per Opening	Opng	-	500.00
	Add for Stainless Steel	Opng	-	420.00
	Add for 30" Diameter	Opng	21.00	200.00
	Deduct for Galvanized	Opng	-	87.00
1004.0	**COAT/ HAT RACKS & CLOSET ACCESS. (M) (Carpenters)**			
	ALUMINUM	LnFt	13.50	80.00
	STAINLESS STEEL	LnFt	13.50	100.00
1005.0	**CUBICLE CURTAIN TRACK & CURTAINS (M) (Carpenters)**	LnFt	8.90	17.50

		UNIT	LABOR	MATERIAL
1006.0	**DIRECTORIES & BULLETIN BOARDS (M) (Carpenters)**			
	(Directories For Changeable Letters)			
	Interior - 48" x 36" Aluminum	Each	145.00	990.00
	Stainless Steel	Each	150.00	1,150.00
	Bronze	Each	150.00	1,600.00
	60" x 36" Aluminum	Each	170.00	1,150.00
	Stainless Steel	Each	175.00	1,700.00
	Bronze	Each	175.00	2,500.00
	Exterior - 48" x 36" Aluminum	Each	200.00	1,380.00
	Stainless Steel	Each	210.00	2,000.00
	Bronze	Each	210.00	2,650.00
	60" x 36" Aluminum	Each	220.00	1,900.00
	Stainless Steel	Each	220.00	2,550.00
	Bronze	Each	220.00	3,370.00
	Illuminated - 72" x 48" Aluminum	Each	220.00	6,000.00
	Stainless Steel	Each	325.00	9,400.00
	Bronze	Each	325.00	11,600.00
1007.0	**DISPLAY & TROPHY CASES (M) (Carpenters)**			
	Aluminum Frame 8' x 6' x 1'6"	Each	335.00	3,580.00
	4' x 6' x 1'6"	Each	200.00	1,950.00
1008.0	**FLOORS - ACCESS OR PEDESTAL (L&M) (Carpenters)**			
	(Includes Laminate Topping)			
	Aluminum	SqFt		20.50
	Steel - Galvanized	SqFt		14.00
	Plywood - Metal Covered	SqFt		13.00
	Add for Carpeting	SqFt		3.00
1009.0	**FIREPLACE AND STOVES**			
1010.0	**FIRE FIGHTING DEVICES (M) (Carpenters)**			
	Extinguishers - 2½ Gal - Water Pressure	Each	25.00	82.00
	Carbon Dioxide - 5 Lb	Each	25.00	190.00
	10 Lb	Each	27.00	265.00
	5 Lb	Each	26.00	60.00
	Dry Chemical - 10 Lb	Each	26.00	74.00
	20 Lb	Each	28.00	120.00
	Halon - 2 1/2 Lb	Each	25.00	132.00
	5 Lb	Each	26.00	158.00
	Cabinets - Enameled Steel - 18 Ga			
	12" x 27" x 8" - Recessed - Portable Ext	Each	45.00	74.00
	24" x 40" x 8" - Recessed - Hose Rack with Nozzle	Each	85.00	152.00
	Deduct for Surface Plastic	Each	-	21.00
	Add for Aluminum	Each	-	67.00
	Blanket & Cabinet - Enameled and Roll Type	Each	45.00	84.00
1011.0	**FLAG POLES (M) (Carpenters)**			
	Aluminum Tapered - Ground 20' - 5" Butt		190.00	1,100.00
	25' - 5 1/2" Butt		200.00	1,550.00
	30' - 6" Butt		260.00	1,680.00
	35' - 7" Butt		345.00	2,150.00
	40' - 8" Butt		470.00	3,500.00
	50' - 10" Butt		590.00	4,800.00
	60' - 12" Butt		760.00	8,200.00
	Wall 15' - 4" Butt		200.00	970.00
	Add for Architectural Grade			25%
	Fiberglass Tapered - Deduct from Above Prices			20%
	Steel Tapered - Deduct from Above Prices			20%
	Steel Tube - Deduct from Above Prices			40%
	Add for Concrete Bases		340.00	230.00

1012.0 FLOOR & ROOF ACCESS DOORS (M) (Carpenters)

	UNIT	LABOR	MATERIAL
Floor - Single Leaf 1/4" Steel			
24" x 24"	Each	97.00	610.00
30" x 30"	Each	123.00	650.00
30" x 36"	Each	152.00	750.00
36" x 36"	Each	172.00	785.00
Deduct for Recess Type	Each		35.00
Add for Aluminum	Each		65.00
Roof - Steel			
30" x 36"	Each	157.00	560.00
30" x 54"	Each	167.00	800.00
30" x 96"	Each	240.00	1,450.00
Add for Disappearing Stairs	Each	740.00	5,000.00
Add for Fixed stairs	Each	210.00	1,350.00

1013.0 FLOOR MATS & FRAMES (M) (Carpenters)

	UNIT	LABOR	MATERIAL
Mats - Vinyl Link 4' x 5' x 1/2"	Each	42.00	370.00
4' x 7' x 1/2"	Each	45.00	550.00
Add for Color	SqFt		4.00
Frames - Aluminum 4' x 5'	Each	78.00	225.00
4' x 7'	Each	84.00	280.00

1014.0 FOLDING GATES (M) (Carpenters)

	UNIT	LABOR	MATERIAL
8' x 8' - with Lock Cylinder	Each	295.00	1,400.00
	or SqFt	4.60	21.90

1015.0 LOCKERS (Incl. Benches) (M)
(Sheet Metal Workers)

	UNIT	LABOR	MATERIAL 60"	MATERIAL 72"
Single - Open Type with 6" Legs				
9" x 12" x 72"	Each	21.00	120.00	125.00
9" x 15" x 72"	Each	21.00	125.00	132.00
12" x 12" x 72"	Each	24.00	135.00	143.00
12" x 15" x 72"	Each	24.50	140.00	144.00
12" x 18" x 72"	Each	25.00	142.00	145.00
15" x 15" x 72"	Each	25.50	144.00	148.00
15" x 18" x 72"	Each	28.00	147.00	152.00
15" x 21" x 72"	Each	29.00	158.00	163.00
Double Tier				
12" x 12" x 36"	Each	16.25	71.00	
12" x 15" x 36"	Each	16.80	74.00	
12" x 18" x 36"	Each	17.10	76.00	
15" x 15" x 36"	Each	17.10	79.00	
15" x 18" x 36"	Each	17.60	86.00	
15" x 21" x 36"	Each	18.10	91.00	
Box Lockers				
12" x 12" x 12"	Each	15.25	44.00	
15" x 15" x 12"	Each	15.75	46.00	
18" x 18" x 12"	Each	16.75	50.00	
12" x 12" x 24"	Each	16.75	60.00	
15" x 15" x 24"	Each	17.25	68.00	
18" x 18" x 24"	Each	17.75	70.00	
Add Flat Key Locks (Master Keyed)	Each	-	8.90	
Add for Combination Locks	Each	-	11.00	
Add for Sloped Tops	Each	7.80	11.00	
Add for Closed Base	Each	7.80	8.50	
Add for Other than Standard Colors	Each	-	15%	
Add for Benches	LnFt	8.80	30.00	
Add for Less than Banks of 4	Each	5%	10%	
Add for Recessed Latch (Quiet)	Each	-	3.50	

1016.0 LOUVERS AND VENTS (M) (Carpenters or Sheet Metal Workers)

	UNIT	LABOR	MATERIAL
LOUVRES			
Exterior			
Galvanized - 16 Ga	SqFt	14.80	44.00
Aluminum - 14 Ga	SqFt	14.80	53.00
Extruded 12 Ga	SqFt	14.80	61.00
Add for Brass, Bronze or Stainless	SqFt	-	50.00
Add for Insect Screen	SqFt	-	4.30
Interior Door or Partition - Steel	SqFt	13.60	55.00
Aluminum	SqFt	13.60	58.00
VENTS - Brick - Aluminum	Each	27.00	73.00
Block - Aluminum	Each	27.00	110.00
See Div. 15 for Mechanically Connected			

1017.0 PARTITIONS - WIRE MESH (M) (Carpenters)

	UNIT	LABOR	MATERIAL
PARTITIONS			
6 Ga - 2" Mesh & Channel Frame	SqFt	5.15	11.00
10 Ga - 1 1/2" Mesh - 8'	SqFt	4.20	8.80
Add for Door	Each	103.00	410.00
WINDOW GUARDS - 10 Ga with Frame	SqFt	6.00	17.50
Add for Galvanized	-	-	15%

1018.0 PARTITIONS AND CUBICLES - DEMOUNTABLE (L&M) (Carpenters)

	UNIT	COST
PARTITIONS - DEMOUNTABLE - 1 9/16" x 10'		
Aluminum or Enameled Steel Extrusions (Acoustical)		
Gypsum Panels (Unpainted)	SqFt	13.80
Vinyl Laminated Gypsum Panels	SqFt	15.20
Steel Panels	SqFt	15.20
Wood Panels - Hard Wood	SqFt	10.00
Add for 2 3/4"	SqFt	2.00
Add per Door	Each	550.00
CUBICLES - DEMOUNTABLE (Includes Glass)		
Steel Frame and Panels (Acoustical) Burlap		
3' - 6" High	SqFt	23.50
4' - 8" High	SqFt	21.30
5' - 8" High	SqFt	18.30
7' - 0" High	SqFt	18.80
Aluminum Frame and Panels (Acoustical) Burlap		
3' - 6" High	SqFt	19.00
4' - 8" High	SqFt	17.70
5' - 8" High	SqFt	17.30
7' - 0" High	SqFt	16.80
Deduct for Non-Acoustical Type	SqFt	2.10
Add for Vinyl Laminated	SqFt	1.90
Add for Carpeted	SqFt	4.25
Add for Curved Panels	SqFt	7.50
Add per Door	Each	450.00
Add per Gate	Each	280.00

1019.0 PARTITIONS - OPERABLE (Including Moving Dividers and Operable Walls) (L&M) (Carpenters)

	UNIT	LABOR	MATERIAL
ACCORDION FOLD (Fabric)	SqFt	5.25	15.80
Add for Acoustical Type	SqFt	-	5.50
Add for Metal & Wood Supports	LnFt	4.75	6.75
FOLDING (Wood, Plastic Lam., Fabric Cover)	SqFt	6.30	32.50
Add for Acoustical Type	SqFt	-	6.10

		UNIT	LABOR	MATERIAL
1019.0	**PARTITIONS - OPERABLE, Cont'd...**			
	OPERABLE WALLS (Electrically Operated)			
	Folding - Plastic Laminated	SqFt	7.50	42.00
	Vinyl Overlaid	SqFt	7.50	40.00
	Wood (Prefinished)	SqFt	7.50	40.00
	Side Coiling - Wood	SqFt	-	55.00
	Metal	SqFt	-	60.00
	REMOVABLE OR PORTABLE WALLS (Acoustical)			
	4"	SqFt	-	56.00
	2 1/4"	SqFt	-	45.00
1020.0	**PEDESTRIAN CONTROLS - GATES & TURNSTILES**	-	-	-
1021.0	**POSTAL SPECIALTIES (M) (Carpenters)**			
	CHUTES - Aluminum	Per Floor	245.00	1,050.00
	Bronze	Per Floor	245.00	1,500.00
	RECEIVING BOX - Aluminum	Each	143.00	990.00
	Bronze	Each	143.00	1,650.00
	LETTER BOXES			
	3" x 5" Alum. with Cam Locks Front Load	Each	11.50	67.00
	Rear Load	Each	11.50	82.00
1022.0	**SHOWER & TUB RECEPTORS & DOORS (M) (Cement Finishers)**			
	PRECAST TERRAZZO WITH STAINLESS STEEL CAP			
	32" x 32" x 6"	Each	108.00	340.00
	36" x 36" x 6"	Each	118.00	420.00
	36" x 24" x 12"	Each	123.00	600.00
1023.0	**SIGNS, LETTERS & PLAQUES (M) (Carpenters)**			
	LETTERS			
	Cast Aluminum, Baked Enamel, Anodized Finish			
	6" x 1/2" x 2"	Each	12.60	30.00
	8" x 1/2" x 2"	Each	13.70	36.00
	12" x 1/2" x 2"	Each	14.75	67.00
	Plastic - Character	Each	3.00	4.20
	PLAQUES - 24" x 24" Cast Aluminum	Each	113.00	970.00
	24" x 24" Bronze	Each	113.00	1,150.00
	SIGNS - HANDICAP (Metal)	Each	14.75	40.00
1024.0	**STORAGE SHELVING (L&M) (Sheet Metal Workers)**			
	METAL - 18 Ga - 36" Sections 7'1" High			
	12" Deep, 6 Shelves	LnFt	15.70	43.00
	18" Deep, 6 Shelves	LnFt	16.70	50.00
	24" Deep, 6 Shelves	LnFt	18.80	59.00
	Add or Deduct per Shelf	LnFt	3.00	5.40
	Add for Closed Back and Ends	LnFt	5.00	5.40
	Add for 48' Sections	LnFt	5.00	3.80
	Deduct for 22 Ga	LnFt	.60	3.80
	PALLET RACKS - 6' x 36" - 4 Shelves	LnFt	31.50	115.00
1025.0	**SUN CONTROL & PROTECTIVE COVERS (L&M)**	SqFt	5.75	21.00
1026.0	**TELEPHONE AND SOUND BOOTHS (M) (Carpenters)**	Each	240.00	3,700.00
1027.0	**TOILET & BATH ACCESSORIES (L&M) (Plumbers)**			
	BASED ON STAINLESS STEEL	UNIT	LABOR	MATERIAL
	Curtain Rods - 60"	Each	20.50	35.00
	Medicine Cabinets - 14" x 24"	Each	34.00	84.00
	18" x 24"	Each	36.00	125.00
	Mirrors - 16" x 10"	Each	22.00	77.00
	18" x 24"	Each	23.00	95.00
	24" x 30"	Each	35.00	115.00
	Add for Shelves	-	-	30.00
	Mop Holders	Each	29.00	80.00
	Paper Dispenser - Surface Mounted	Each	23.00	36.00
	Recessed	Each	29.00	70.00
	Toilet Seat Covers - Recessed	Each	29.00	82.00

		UNIT	LABOR	MATERIAL
1027.0	**TOILET & BATH ACCESSORIES, Cont'd...**			
	Purse Shelf	Each	24.50	70.00
	Robe Hooks	Each	18.00	17.50
	Handicap Grab Bar - 18"	Each	24.00	41.00
	36"	Each	27.00	48.00
	42"	Each	33.50	53.00
	Set of Three	Each	73.00	107.00
	Soap Dispenser - Wall Mount	Each	24.00	52.00
	- with Lather Shelf	Each	25.00	107.00
	Soap and Grab Bar	Each	24.00	59.00
	Toilet Seat Cover Dispenser	Each	27.00	82.00
	Towel Bars - 24"	Each	24.00	35.00
	Towel Dispenser - Recessed	Each	50.00	470.00
	- Surface Mounted	Each	46.00	150.00
	and Disposal - Recessed	Each	46.00	470.00
	Sanitary Napkin Disposal - Recessed	Each	26.00	70.00
	Electric Hand Dryer	Each	68.00	385.00
1028.0	**TOILET COMPARTMENTS (Incl. Shower & Dressing Rooms) (L&M) (Sheet Metal Workers & Plumbers)**			
	PARTITIONS - FLOOR BRACED			
	(Door & 1 Side Panel) Baked Enameled	Each	130.00	450.00
	Stainless Steel	Each	135.00	1,070.00
	Plastic Laminate	Each	140.00	500.00
	Marble	Each	220.00	830.00
	Add for Extra Side Panel - Baked Enameled	Each	78.00	225.00
	Plastic Laminate	Each	78.00	250.00
	Marble	Each	158.00	320.00
	Add for Ceiling Hung	Each	78.00	107.00
	Add for Reinforcing for Handicap Bars (ADA)	Each	-	80.00
	Add for Paper Holders	Each	16.50	38.00
	Urinal Screens - Baked Enameled	Each	95.00	225.00
	Plastic Laminate	Each	95.00	240.00
	Marble	Each	138.00	410.00
	Shower Partitions (No Plumbing)			
	Baked Enameled (with Curtain & Terrazzo Base)			
	32" x 32"	Each	168.00	700.00
	36" x 36"	Each	168.00	640.00
	Fiberglass & Fiberglass Base - 32" x 32"	Each	105.00	290.00
	36" x 36"	Each	105.00	330.00
	Add for Doors	Each	45.00	225.00
	Tub Enclosures			
	Aluminum Frame and Tempered Glass Door	Each	78.00	210.00
	Chrome Frame and Tempered Glass Door	Each	78.00	275.00
1029.0	**URNS AND TRAYS**			
	CIGARETTE URNS - SS Rectangular or Round	Each	44.00	108.00
1030.0	**WALL AND CORNER GUARDS**			
	CORNER GUARDS			
	Stainless Steel			
	16 Ga 4" x 4" with Anchor	LnFt	6.80	22.00
	18 Ga 4" x 4" with Anchor	LnFt	6.30	22.00
	22 Ga 3½" x 3½" with Anchor	LnFt	6.80	18.00
	22 Ga 3½" x 3½" with Tape	LnFt	6.30	17.00
	Aluminum - Anodic with Tape	LnFt	6.30	15.00
	Vinyl	LnFt	5.75	9.60
	See 0502.3 for Steel			

		UNIT	LABOR	MATERIAL
1101.0	**MEDICAL CASEWORK (L&M) (Carpenters)**			
	INCLUDING PLASTIC LAMINATE TOP, HARDWARE & FINISH			
.1	METAL			
	Base Cabinets - 35" High x 24" Deep	LnFt	33.50	210.00
	Wall Cabinets - 24" High x 12" Deep	LnFt	31.50	165.00
	High Wall (Utility) - 84" High x 12" Deep	LnFt	34.50	170.00
.2	PLASTIC LAMINATE (Including Plastic Laminator)			
	Base Cabinets - 35" High x 24" Deep	LnFt	33.50	220.00
	Wall Cabinets - 24" High x 12" Deep	LnFt	31.50	155.00
	High Wall (Utility) - 84" High x 12" Deep	LnFt	34.50	175.00
.3	WOOD			
	Base Cabinets - 35" High x 24" Deep	LnFt	33.50	205.00
	Wall Cabinets - 24" High x 12" Deep	LnFt	31.50	155.00
	High Wall (Utility) - 84" High x 12" Deep	LnFt	34.50	175.00
.4	STAINLESS STEEL			
	Base - 35" High x 24" Deep	LnFt	34.50	245.00
	Wall - 24" High x 12" Deep	LnFt	31.50	225.00
1102.0	**EDUCATIONAL CASEWORK (L&M) (Carpenters)**			
	Same as 1101.0 Medical Casework above.			
1103.0	**KITCHEN AND BATH CASEWORK (M) (Carpenters)**			
.1	PREFINISHED WOOD - NO TOPS			
	Base Cabinets - 35" High x 18" Deep	LnFt	33.50	175.00
	Front Only	LnFt	31.50	100.00
	with 1 Drawer	LnFt	33.50	210.00
	with 4 Drawers	LnFt	34.50	280.00
	with Lazy Susan	LnFt	33.50	220.00
	Upper Cabinets - 30" High x 12" Deep	LnFt	32.50	140.00
	18" High x 12" Deep	LnFt	31.50	110.00
	with Lazy Susan	LnFt	32.50	185.00
	High Cabinets - 84" High x 18" Deep	LnFt	34.50	280.00
	Add for Counter Tops - See 0603.2			
.2	PLASTIC LAMINATE			Add 10%
1104.0	**MEDICAL EQUIPMENT (M) (Carpenters)**			
.1	STATION AND SERVICE CENTER EQUIPMENT			
.11	Kitchen -			
	Example: 4'6" x 4'2" Base & 30" Upper	Each	240.00	2,000.00
.12	Medical Preparation -			
	Example: 4'0" x 6'8" x 1'8" (also 5' and 6' wide)	Each	300.00	4,800.00
.13	Nourishment and Ice Stations -			
	Example: 6'0" x 6'8" x 2'8"	Each	320.00	10,500.00
.14	Lavatory Unit -			
	Example: 4'7" x 1'8" x 4'0"	Each	150.00	1,000.00
.15	Janitor's Closet -			
	Example: 2'0" x 5'4" x 6'8"	Each	275.00	4,000.00
.16	Glove Packaging - Example	Each	150.00	1,100.00
.17	Linen Inspection - Example	Each	170.00	1,550.00
.2	LAB TABLES AND DEMONSTRATION DESKS	Each	160.00	2,300.00

1104.0 MEDICAL EQUIPMENT, Cont'd...

		UNIT	LABOR	MATERIAL
.3	TOPS AND BOWLS (M) (Carpenter or Plumber)			
	Calcium Aluminum Silicate	SqFt	6.30	24.70
	Corian	SqFt	6.30	47.00
	Epoxy Resins (Kemstone)	SqFt	6.30	23.00
	Marble - Artificial	SqFt	5.00	25.00
	Plastic Laminate - 25" - 1 1/2"	SqFt	4.20	23.00
	Add for Backsplash	LnFt	1.80	4.60
	Soapstone	SqFt	4.70	35.00
	Stainless Steel - Bowl	Each	68.00	198.00
	Tops	SqFt	6.30	64.00
	Add for Backsplash	LnFt	2.10	8.25
	Vinyl Sheet - 25" - 1 1/2"	SqFt	3.15	14.00
	Add for Backsplash	LnFt	1.75	4.65
.4	FUME HOODS (INCLUDING DUCT WORK)			
	Stock - 4"	Each	430.00	5,250.00
	Custom	Each	430.00	5,800.00
.5	KEY CABINETS (100 KEY)	Each	84.00	310.00
.6	BLANKET AND SOLUTION WARMERS (NO MECHANICAL OR ELECTRICAL)			
	Example: 27 x 11 x 30" x 74			
	Free Standing - Steam	Each	395.00	5,000.00
	Electric	Each	395.00	5,200.00
	Recessed - Steam	Each	445.00	4,900.00
	Electric	Each	445.00	5,000.00
.7	ENVIRONMENTAL GROWTH CHAMBERS & LABS	Each	1,200.00	4,750.00
.8	CHART RACKS AND CHART RACK DESKS			
	Racks	Each	73.00	330.00
	Desks	Each	110.00	1,600.00

			ELECTRIC		STEAM	
.9	STERILIZING EQUIPMENT (NO MECHANICAL/ ELECTRICAL)	UNIT	Re-cessed	Open Mtd or Mobile	Re-cessed	Open Mtd or Mobile
	Dressing and Instrument -					
	Example: 16" x 26"	Each	10,800	-	1,000	-
	Dressing -					
	Example: 20" x 20" x 38"	Each	22,500	-	17,500	1,800
	Solution Warming/ Storage Cab.	Each	-	-	-	-
	Bedpan -					
	Example: 16" x 16" x 27"					
	Non Pressure	Each	-	2,500	-	-
	Pressure	Each	-	3,000	-	-
	Laboratory (Painted)	Each	12,500	13,250	11,000	13,750
	Add for Stainless Steel					
	Autoclave (All Purpose)					
	Portable Inst. - Non Pressure					
	6" x 6" x 13"	Each	-	1,080	-	-
	8" x 8" x 16"	Each	-	1,700	-	-
	10" x 10" x 22"	Each	-	5,500	-	-
	Hot Air -					
	Example: 39" x 14" x 19"					
	Gravity	Each	2,900	2,650	-	-
	Convection	Each	4,200	5,000	-	-

1104.0 MEDICAL EQUIPMENT, Cont'd...

.10 X-RAY AND DARK ROOM EQUIPMENT

Lead Protection - See 0801 for Doors & View Window Frames

Lead Protection - See 0902 for Lead Lined Sheetrock

Manual Dark Room

Item	UNIT	LABOR	MATERIAL
Example: Room including Cassette Pass Box, Cassette Storage Cabinet, Film Identifier Cabinet, Film Illuminator and Dryer, Development Tank including Tops and Splashes	Each	1,580.00	15,500.00
Automatic			
Processor with Solution Storage Tank	Each	840.00	11,700.00
Film Loader	Each	295.00	5,750.00
.11 AUTOPSY AND MORTUARY EQUIPMENT			
Refrigerators	Each	480.00	9,800.00
Tables	Each	315.00	6,100.00
Carts	Each	87.00	1,625.00
.12 NARCOTIC SAFES	Each	98.00	630.00
.13 REFRIGERATORS (OTHER THAN MORTUARY)			
Under Counter	Each	240.00	2,600.00
.14 TOTE TRAYS AND UTILITY CARTS	Each	84.00	500.00
.15 GAS TRACKS - Example: Operating Room	Each	520.00	3,400.00
Straight	LnFt	46.00	285.00
Curved	LnFt	57.00	345.00
.16 ANIMAL FENCING AND CAGING	SqFt	2.70	6.30
.17 DENTAL	-	-	-
.18 SURGICAL	-	-	-
.19 INCUBATORS	Each	-	4,700.00
.20 THERAPY-HEAT	Each	-	2,250.00
.21 OVERHEAD HOISTS	Each	-	2,200.00

1105.0 EDUCATIONAL EQUIPMENT

1106.0 RESIDENTIAL EQUIPMENT (NO MECH. OR ELEC.)

See 1504 for Supply, Waste and Vent

See 1611 for Electric Power

Item	UNIT	LABOR	MATERIAL
.1 FREE STANDING			
Clothes Dryer - Gas	Each	87.00	490.00
Electric	Each	87.00	440.00
Clothes Washers	Each	87.00	460.00
Compactors - 15"	Each	83.00	520.00
Disposals	Each	83.00	175.00
Dishwasher	Each	87.00	510.00
Freezers - Upright - 16 CuFt	Each	87.00	460.00
19 CuFt	Each	94.00	500.00
21 CuFt	Each	105.00	550.00
Freezers - Upright - Frost Proof			
15 CuFt	Each	87.00	565.00
18 CuFt	Each	87.00	620.00
Freezers - Chest - 15 CuFt	Each	87.00	515.00
20 CuFt	Each	93.00	620.00
25 CuFt	Each	105.00	715.00
Dehumidifier - 40 Pint	Each	87.00	190.00
Humidifier - 10 Gals/Day	Each	87.00	265.00
Range Hood	Each	72.00	185.00
Water Heater - 30 Gal. & 40 Gal.	Each	130.00	240.00
Water Softener - 30 Grains	Each	157.00	690.00

1106.0 RESIDENTIAL EQUIPMENT, Cont'd...

		UNIT	LABOR	MATERIAL
.1	FREE STANDING, Cont'd.			
	Refrigerators			
	Compact - 11 CuFt	Each	88.00	370.00
	Conventional- 14 CuFt	Each	95.00	500.00
	With Freezer- 15 CuFt (Also 17 & 19)	Each	100.00	620.00
	21 CuFt	Each	105.00	770.00
	Frostproof- 12 CuFt	Each	95.00	640.00
	15 CuFt	Each	107.00	840.00
	Side-by-Side- 22 CuFt	Each	107.00	1,190.00
	25 CuFt	Each	110.00	1,350.00
	Ranges - Electric - 30"	Each	88.00	690.00
	40"	Each	94.00	1,410.00
	Gas - 20"	Each	83.00	305.00
	30"	Each	89.00	690.00
	Microwave Ovens - Counter	Each	89.00	200.00
	Over-counter	Each	125.00	460.00
.2	BUILT-IN			
	Air Conditioners - 110 Volt	Each	83.00	390.00
	220 Volt	Each	83.00	590.00
	Dishwashers	Each	83.00	500.00
	Disposals	Each	83.00	172.00
	Hoods	Each	83.00	190.00
	Ranges - Surface - 30" Gas	Each	83.00	305.00
	42" Gas with Broiler	Each	83.00	1,350.00
	30" Electric	Each	83.00	315.00
	42" Elec. with Broiler	Each	88.00	1,350.00
	Ovens - Single - Gas	Each	88.00	465.00
	Double	Each	100.00	590.00
	Single - Electric	Each	100.00	420.00
	Double	Each	100.00	525.00

1107.0 ATHLETIC EQUIPMENT (M) (Carpenters)

		UNIT	LABOR	MATERIAL
.1	BASKETBALL BACKSTOPS			
	Stationary	Each	390.00	715.00
	Retractable	Each	550.00	1,750.00
	Suspended	Each	660.00	3,500.00
	Add for Electric Operated	Each	-	1,040.00
.2	GYM SEATING - See Div. 1306.0			
.3	SCOREBOARDS (Great Variable)	Each	1050.00	6,300.00
.4	BOWLING AND BILLIARDS - See Div. 1302.0			
.5	SWIMMING POOL EQUIPMENT- See Div. 1318.0			
.6	OUTSIDE RECREATIONAL EQ.- See Div. 0208.2			
.7	GYMNASTIC EQUIPMENT - See Div. 0208.2			
.8	SHOOTING RANGE EQUIPMENT	Point	780.00	6,700.00
.9	HEALTH CLUB EQUIPMENT	-	-	-

1108.0 FOOD SERVICE EQUIPMENT (L&M) (Sheet Metal and Plumbing)

	UNIT		COST
Broilers - 36" Electric	Each	-	5,000.00
Can Washers	Each	-	3,600.00
Coffee Stand and Urn	Each	-	4,350.00
Cold Food Carts	Each	-	4,750.00
Conveyors - 12'	Each	-	6,400.00
Cutter - Mixer	Each	-	7,500.00
Dishtables - Clean	Each	-	4,000.00
Soiled	Each	-	5,300.00

		UNIT		COST
1108.0	**FOOD SERVICE EQUIPMENT, Cont'd...**			
	Disposer - Garbage - 3 H.P.	Each		4,100.00
	1 1/2 H.P.	Each		2,300.00
	Dishwasher- Automatic	Each		19,500.00
	Rack Type	Each		6,900.00
	Fryers - Double	Each		2,400.00
	Hot Food Carts	Each		3,750.00
	Ice Cream Dispenser	Each		1,420.00
	Ice Machine with Bin	Each		2,000.00
	Juice and Beverage Dispenser	Each		2,100.00
	Meat Saw	Each		3,600.00
	Microwave	Each		365.00
	Oven - Double Convector - Electric	Each		7,250.00
	Single	Each		3,700.00
	Portable Sink	Each		950.00
	Range - Oven- Four Burner	Each		1,475.00
	Range Hoods	Each		15,100.00
	Refrigerators	Each		2,500.00
	Silverware Dispensers	Each		240.00
	Sinks - Utility	Each		1,325.00
	Soiled Dish	Each		4,500.00
	Slicer	Each		2,600.00
	Steamers	Each		5,250.00
	Tray Dispensers	Each		740.00
	Setup Unit	Each		1,400.00
	Utility Carts	Each		225.00
	Water Stations	Each		1,130.00
	Work Tables	Each		1,200.00
	Average per Kitchen for Equipment	SqFt		68.00
1109.0	**LAUNDRY EQUIPMENT (L&M)**			
	Dryer - 50 Lb.	Each		2,100.00
	Ironer, Air	Each		7,400.00
	Hand	Each		5,100.00
	Washer Tumbler - 135 Lb.	Each		36,000.00
	50 Lb.	Each		20,000.00
1110.0	**LIBRARY EQUIPMENT (M) (Carpenters)**			
			LABOR	MATERIAL
	Book Trucks	Each	100.00	1,250.00
	Card Catalog Index - 30 drawer	Each	127.00	2,200.00
	Chairs	Each	17.80	115.00
	Charging Desks - Card File	Each	100.00	1,100.00
	Charging Unit	Each	100.00	1,180.00
	Open Shelf	Each	100.00	780.00
	Newspaper Rack	Each	95.00	760.00
	Shelving (Maple) - Example: 36" x 60" x 8"	LnFt	17.80	115.00
	Tables	Each	73.00	770.00
1111.0	**ECCLESIASTICAL EQUIPMENT (M) (Carpenters)**			
.1	CROSSES, SPIRES AND STEEPLES	-	-	-
.2	CONFESSIONALS - Two Person	Each	600.00	5,900.00
	One Person	Each	480.00	4,100.00
.3	PEWS (OAK) (With Kneelers)	LnFt	15.70	115.00
	Add for Frontals	LnFt	15.70	58.00
	Add for Padded Kneelers	LnFt	4.20	3.40
	Add for Birch	LnFt	-	12.00
.4	BAPTISMAL FONTS	Each	70.00	830.00
.5	MISCELLANEOUS (WOOD ONLY)			
	Main Altars	Each	470.00	5,850.00
	Side Altars	Each	260.00	2,200.00
	Lecterns	Each	157.00	1,300.00
	Credence Table	Each	132.00	850.00

		UNIT	LABOR	MATERIAL
1112.0	**STAGE AND THEATRICAL EQUIPMENT (L&M) (Carpenters)**			
.1	STAGE DRAPERIES, HARDWARE, CONTROLS AND OTHER EQUIPMENT			
	Motorized Rolling Curtain - 38' x 16'	Each	-	6,250.00
	Stage Equipment Sets	Each	-	2,200.00
	Light Bridge	Each	-	22,000.00
.2	FOLDING STAGES - Portable Thrust	Each	-	14,500.00
.3	REVOLVING STAGES - 32' Dia x 12" High	Each	-	135,000.00
.4	FIXED AUDITORIUM SEATING - Cushioned	Each	-	195.00
.5	PROJECTION SCREENS			
	Motor Oper. - 8' x 8'	Each	137.00	1,600.00
	10' x 10'	Each	142.00	1,900.00
	Stationary - 20' x 20'	Each	185.00	3,050.00
1113.0	**BANK EQUIPMENT (M) (Carpenters)**			
.1	SAFES	Each	167.00	1,200.00
.2	DRIVE-UP WINDOW UNIT - Including Cash Drawers, Counter, Window, Depository and Lock Box	Each	1,000.00	9,200.00
.3	TELLER UNIT	Each	340.00	2,500.00
	Cash Dispensing	Each	3,150.00	42,000.00
.4	NIGHT DEPOSITORY - Envelope	Each	230.00	1,150.00
	Incl. Head & Chest	Each	1,030.00	10,400.00
.5	SURVEILLANCE SYSTEMS			
	Receiver and Transmitter	Each	-	750.00
	Cameras	Each	-	1,100.00
	Time Lapse Recorder	Each	-	3,600.00
.6	SAFE DEPOSIT BOX	Each	7.90	80.00
	Vault Doors - See Div. 0803.16			
	Pneumatic Tube System - See Div. 14			
1114.0	**DETENTION EQUIPMENT**			
1115.0	**PARKING EQUIPMENT**			
.1	AUTOMATIC DEVICES			
	Cash Registers	Each	575.00	12,400.00
	Clocks - Rate Computing	Each	90.00	900.00
	Standards	Each	73.00	700.00
	Gate	Each	420.00	3,400.00
	Loop Sensor	Each	230.00	1,020.00
	Treadle	Each	175.00	420.00
	Operator Station - Coin	Each	525.00	5,100.00
	Key	Each	530.00	1,100.00
	Radio Control - Transmitter	Each	-	2,000.00
	Ticket Spitter Station	Each	262.00	6,250.00
.2	BOOTHS - Example: 4' x 6'	Each	315.00	6,100.00
.3	METERS AND POSTS	Each	89.00	335.00
1116.0	**LOADING DOCK EQUIPMENT**			
.1	DOCK BUMPERS			
	16" x 12" x 4"	Each	24.00	53.00
	24" x 6" x 4"	Each	24.00	54.00
	24" x 10" x 4"	Each	25.00	58.00
	24" x 12" x 4"	Each	26.00	79.00
	36" x 6" x 4"	Each	27.00	70.00
	36" x 10" x 4"	Each	29.00	87.00
	36" x 12" x 4"	Each	33.00	102.00
	Add for Galvanized Angles	Each	-	7.90

		UNIT	LABOR	MATERIAL
1116.0	**LOADING DOCK EQUIPMENT, Cont'd...**			
.2	DOCK BOARDS -			
	5,000# 3' - 6" x 6' - 0"	Each	122.00	1,020.00
	10,000# 4' - 0" x 6' - 0"	Each	137.00	1,300.00
	15,000# 4' - 0" x 6' - 0"	Each	147.00	1,720.00
.3	LEVELERS & ADJUSTABLE RAMPS			
	8' - 0" x 8' (Add Concrete)	Each	240.00	3,200.00
.4	SHELTERS - 10' x 10' with Dock Pad	Each	340.00	1,950.00
	Strip Curtain - 10' x 10' x 8"	Each	95.00	255.00
	Dock Pad - 10' x 10'	Each	190.00	700.00
1117.0	**SERVICE STATION EQUIPMENT**			
.1	REELS - Chassis (No Mech. or Electric)	Each	132.00	680.00
	Air	Each	132.00	570.00
	Water	Each	132.00	590.00
	Motor Oil	Each	132.00	680.00
	A.T.F.	Each	132.00	740.00
	Gear Lube	Each	132.00	670.00
	Combination of Above - 3 Hose	Each	545.00	5,350.00
	Combination of Above - 5 Hose	Each	890.00	6,600.00
.2	PUMPS - Chassis	Each	157.00	600.00
	Gear Oil	Each	157.00	660.00
	A.T.F.	Each	157.00	660.00
	Oil - Motor	Each	115.00	700.00
	Gasoline	Each	340.00	2,100.00
.3	COMPRESSORS - 5 H.P.	Each	340.00	4,500.00
	3 H.P.	Each	330.00	2,200.00
	1 1/2 H.P.	Each	330.00	2,100.00
.4	TIRE CHANGERS - AUTO	Each	430.00	3,100.00
	Exhaust Systems - See Div. 15			
	Hoisting Equipment - See Div. 14			
	Tanks/Air/Water Connections - See Div. 15			
1118.0	**MUSICAL EQUIPMENT**			
1119.0	**CHECKROOM EQUIPMENT**			
	MOTORIZED, 1-Shelf, 16 ft	LnFt	21.00	120.00
	2-Shelf, 16 ft	LnFt	29.50	245.00
1120.0	**AUDIO VISUAL EQUIPMENT**			
	PROJECTION SCREEN	Each	-	132.00
	ELECTRIC CONTROLLED	Each	-	530.00
	TELEVISION CAMERA	Each	-	1,050.00
	MONITOR	Each	-	750.00
	VIDEO TAPE RECORDER	Each	-	1,080.00
1121.0	**INCINERATORS**			
	SMALL - 18" x 18"	Each	240.00	2,150.00
	MEDIUM - 22" x 22"	Each	260.00	3,150.00
	LARGE	Each	395.00	4,400.00
1122.0	**PHOTO AND GRAPHIC ART EQUIPMENT**			
1123.0	**WASTE HANDLING EQUIPMENT**			
	COMPACTORS - Bag - 3 CuYd Hopper	Each	950.00	11,800.00
	Cart	Each	950.00	10,500.00

1201.0 MANUFACTURED CASEWORK (L&M) (Carpenters)
(Including Plastic Laminate Top & Hardware)

		UNIT	LABOR	MATERIAL
.1	MEDICAL CASEWORK			
.11	Metal			
	Base Cabinets - 35” High x 24” Deep	LnFt	34.70	220.00
	Upper Cabinets - 24” High x 12” Deep	LnFt	33.40	175.00
	High Wall - 84” High x 12” Deep	LnFt	34.70	180.00
.12	Plastic Laminate			
	Base Cabinets - 35” High x 24” Deep	LnFt	34.70	235.00
	Upper Cabinets - 24” High x 12” Deep	LnFt	33.40	170.00
	High Wall - 84” High x 12” Deep	LnFt	34.70	270.00
.13	Wood - Prefinished			
	Base Cabinets - 35” High x 24” Deep	LnFt	34.70	220.00
	Upper Cabinets - 24” High x 12” Deep	LnFt	33.40	165.00
	High Wall - 84” High x 12” Deep	LnFt	34.70	265.00
.14	Stainless Steel			
	Base - 35” High x 24” Deep	LnFt	35.70	270.00
	Wall - 24” High x 12” Deep	LnFt	34.70	245.00
.2	EDUCATIONAL CASEWORK			

		UNIT	COST
1202.0	**BLINDS, SHADES AND SHUTTERS (L&M) (Carpenters)**		
.1	BLINDS - Horizontal		
	Wood 1"	SqFt	6.40
	Aluminum 1"	SqFt	6.00
	1"	SqFt	4.75
	Steel 2"	SqFt	4.40
	Cloth 1"	SqFt	7.00
	Vertical		
	Cloth or PVC 3"	SqFt	5.80
	Aluminum 3"	SqFt	7.35
.2	SHADES - Cotton	SqFt	2.08
	Vinyl Coated	SqFt	2.37
	Lightproof	SqFt	3.05
	Slat	SqFt	2.46
	Woven Aluminum	SqFt	4.30
	Fibre Glass	SqFt	3.00
	X-ray and Dark Room	SqFt	10.50
.3	SHUTTERS		
	16" x 1 1/8" x 48"	Pair	115.00
	16" x 1 1/8" x 60"	Pair	132.00
	16" x 1 1/8" x 72"	Pair	138.00

		UNIT	COST
1203.0	**DRAPERIES AND CURTAIN HARDWARE (L&M) (Carpenters)**		
	TRACK	LnFt	5.15
	DRAPERIES	SqYd	9.20
	Add for Lined	SqYd	2.05
	Add for Rods	SqYd	2.35
	Add for Motorized	SqYd	7.00
1204.0	**OPEN OFFICE FURNITURE**		

		UNIT	LABOR	MATERIAL
1205.0	**FLOOR MATS AND FRAMES (M) (Carpenters)**			
	MATS - Vinyl Link - 4' x 5' x 1/2"	Each	33.40	345.00
	4' x 7' x 1/2"	Each	33.40	485.00
	Add for Color	SqFt	-	17.20
	FRAMES - Aluminum - 4' x 5'	Each	68.00	215.00
	4' x 7'	Each	79.00	275.00

		UNIT	COST
1206.0	**AUDITORIUM AND THEATRE SEATING (L&M) (Carpenters)**		
	WOOD AND PLYWOOD	Each	160.00
	CUSHIONED	Each	190.00
	ROCKING	Each	270.00
1207.0	**MULTIPLE-USE SEATING**		
	MOVABLE - Chrome and Solid Plastic	Each	77.00
	Wood and Plywood	Each	68.00
1208.0	**BUILT-IN FOLDING TABLES AND SEATING (M) (Carpenters)**		
.1	TABLES - 3' x 8' - Tempered Hardboard Top	Each	212.00
	Vinyl Laminate Top	Each	238.00
	Folding Table & Bench-Pocket Unit	Each	1,750.00
.2	CHAIRS - Metal	Each	28.00

1301.0 AIR SUPPORTED STRUCTURES

		UNIT	COST
.1	TENNIS COURTS - 2 LAYER THERMAL DOMES (Includes Doors, Heating, Equipment, Lights & Power)		
	1 Court - 58' x 118'	SqFt	20.00
	2 Courts - 106' x 118'	SqFt	19.00
	3 Courts - 156' x 118'	SqFt	17.00
	4 Courts - 201' x 118'	SqFt	16.00
	Deduct for Non Thermal Type	SqFt	2.20
.2	CONSTRUCTION DOMES, WAREHOUSES, ETC.	SqFt	8.45
	Add for Heating	SqFt	3.70
	Add for Light and Power	SqFt	1.90
	Add for Air Lock Doors	Each	14,700.00
	Add for Revolving Doors	Each	7,600.00

1302.0 BOWLING ALLEYS (L&M) (Carpenters)

	UNIT	COST
BOWLING ALLEYS (L&M) (Carpenters)	Lane	40,000.00
Add for Automatic Scorers	Lane	16,000.00

1303.0 CHIMNEYS - SPECIAL (Foundations Not Included)

.1 JOB CONSTRUCTED (L&M) (Bricklayers)

Inside Diam.	Height		UNIT	COST
5'	100'	Radial Brick	Each	132,000.00
6'	150'	Radial Brick	Each	162,000.00
7'	200'	Radial Brick	Each	212,000.00
8'	250'	Reinforced Concrete	Each	540,000.00
9'	200'	Reinforced Concrete	Each	660,000.00
10'	250'	Reinforced Concrete	Each	880,000.00
12'	300'	Reinforced Concrete	Each	1,500,000.00

.2 PREFABRICATED (METAL REFRACTORY LINED)
1800° - 2000° (L&M) (Sheet Metal & Iron Workers)

	UNIT	COST
10"	LnFt	64.00
12"	LnFt	68.00
15"	LnFt	81.00
18"	LnFt	98.00
21"	LnFt	118.00
24"	LnFt	140.00
30"	LnFt	250.00
36"	LnFt	350.00
Add for Cleanouts or T's	Each	360.00
Add for V.L. Label	Each	10%
Deduct for 800° Chimney	Each	5%

1304.0 CLEAN ROOMS

	UNIT	COST
CLEAN ROOMS	SqFt	200.00 – 400.00

1305.0 FLOORS - ACCESS OR PEDESTAL (L&M) (Carpenters)

(Includes Laminate Topping)

	UNIT	COST
Aluminum	SqFt	21.00
Steel - Galvanized	SqFt	14.00
Plywood - Metal Covered	SqFt	13.00
Add for Carpeting	SqFt	3.35
Add for Stairs and Ramps	SqFt	50%

1306.0 GRANDSTANDS, BLEACHERS AND GYM SEATING

		UNIT	COST
.1	GRANDSTANDS AND BLEACHERS		
	Not Permanent	Each Seat	27.00
	Permanent - Aluminum	Each Seat	105.00
	Steel	Each Seat	68.00
	Concrete	Each Seat	250.00
.2	GYM SEATING - FOLDING	Each Seat	77.00

1307.0 GREENHOUSE (L&M) (Glaziers)

SINGLE ROOM TYPE - Commercial	SqFt	23.50
Educational	SqFt	31.50
Residential	SqFt	34.75
Add for Heating, Lighting and Benches	SqFt	20.00
Add for Cooling	SqFt	8.30

1308.0 INCINERATORS (L&M) (Ironworkers and Bricklayers)

Stack and Breaching Not Included

Pounds/Hour		
100	Each	9,000.00
200	Each	9,100.00
300	Each	10,300.00
400	Each	11,600.00
500	Each	12,800.00
600	Each	16,200.00
800	Each	26,500.00
Add for Gas Burner	Each	1,800.00
Add for Controls	Each	950.00
Add for Piping	Each	1,040.00

1309.0 INSULATED ROOMS (L&M) (Carpenters)

.1	CUSTOM - Cooler	SqFt	185.00
	Freezer	SqFt	215.00
.2	PREFABRICATED (SELF-CONTAINED)		
	Cooler Example: Room 7'6" x 9'6"		
	Wood (Plywood)	Each	7,000.00
	Metal Covered 20 ga Galvanized #3 SS	Each	8,700.00
	Add for 3/4 HP Compressor	Each	1,150.00
	Add for 3/4 HP Compressor Starter & Coils	Each	5,100.00
	Add for Electrical	Each	925.00
	Cooler - Total Cost - Wood	SqFt	195.00
	Metal Covered	SqFt	215.00
	Freezer Example: Room 7'6" x 9'6"		
	Wood (Plywood)	Each	7,300.00
	Metal Covered 20 ga Galvanized #3 SS	Each	9,000.00
	Add for 3/4 HP Compressor	Each	1,500.00
	Add for 3/4 HP Compressor Starter & Coils	Each	4,600.00
	Add for Electrical	Each	950.00
	Freezer - Total Cost - Wood	SqFt	215.00
	Metal Covered	SqFt	240.00
	Add for Shelving	SqFt	9.90

		UNIT	COST
1310.0	**METAL BUILDINGS**		
.1	SHELL - FLOOR AREA (NO FOUNDATION)	SqFt	15.80
	24 ga. 12' Sidewalls and 60' Span (Galvanized)		
	Add per foot above 12' - Floor Area	SqFt	.44
	Add per foot of Span above or below 60'	SqFt	.27
	Add for 2' Overhand - Overhang Area	SqFt	6.10
	Add for Aluminum	SqFt	.70
	Add for Liner Panels - Wall Area	SqFt	2.20
	Add for Insulation 3 1/2" - Wall Area	SqFt	.85
	Add for Insulation 6" - Ceiling Area	SqFt	1.05
	Add for Door Openings - Including Hardware	Each	770.00
	Add for Window Openings	Each	400.00
	Add for Insulated Glass Area	SqFt	11.30
	Add for Overhead Door Area	SqFt	10.70
	See 0803.4 for Door Cost		
	Add for Gutter	LnFt	5.25
	FOUNDATION & GROUND SLAB COST AVG (3-Foot Frost)	SqFt	8.75
1311.0	**RADIATION PROTECTION (L&M) (Carpenters)**		
.1	ELECTROSTATIC SHIELDING		
	Custom - Vinyl Sandwiched Copper Mesh for Floors, Walls & Ceilings (including Copper Foil Tabs & Plastic Tape)	SqFt	26.75
	Add per door and Frame with RF Shielded Glass (including Grounding Wires)	Each	4,000.00
	Add per Borrowed Light RF Shielded Glass	Each	1,650.00
	Example: Room 10' x 8' (Prefab)	Room	32,000.00
		or SqFt	400.00
.2	ELECTROMAGNETIC SHIELDING		
	Example: Room 10' x 8' (Prefab)	Room	13,500.00
		or SqFt	169.00
	Add for Finish, Electricity and Erection	Room	11,200.00
.3	X-RAY		
	Deep Therapy		
	Example: Room 10' x 15'	Room	13,500.00
	Radiography or Fluoroscopy		
	Example: Room 10' x 15'	Room	5,800.00
.4	RADIO FREQUENCY SHIELDING		
	Example: Room 10' x 10'	Room	33,600.00
	See 1104.10 for X-Ray Equipment		
1312.0	**SAUNAS (M) (Carpenters)**		
	Examples (includes Heaters):		
	Room 5' x 6' Cedar (5 K.W. Unit)	Room	5,200.00
		or SqFt	173.50
	Room 6' x 6' Cedar (6 K.W. Unit)	Room	6,400.00
		or SqFt	178.00
	Room 6' x 7' Cedar (7.5 K.W. Unit)	Room	6,400.00
		or SqFt	152.50
	Add for Electrical Connections and Ventilating		900.00

1313.0 SCALES (NO PORTABLE) (Ironworkers)

.1 FLOOR SCALES (HEAVY DUTY INDUSTRIAL)

Size	Capacity	Labor	Material	Concrete Work
46" x 38"	800 lb.	460.00	4,800.00	-
48" x 48"	1,350 lb.	490.00	5,100.00	-
60" x 48"	1,600 lb.	690.00	5,500.00	-
72" x 48"	2,600 lb.	715.00	6,400.00	-
76" x 54"	6,000 lb.	1,250.00	6,700.00	-
84" x 60"	10,000 lb.	2,050.00	8,800.00	-

.2 TRUCK SCALES - FULL WITH WEIGH BEAM

Truck (4-Section) -				
45" x 10"	50 Ton	2,950.00	23,900.00	22,500.00
50" x 10"	50 Ton	3,250.00	24,500.00	23,000.00
60" x 10"	50 Ton	3,700.00	25,700.00	23,500.00
70" x 10"	50 Ton	4,400.00	26,700.00	25,200.00
Add for 60-Ton Capacity			3,800.00	
Axle Load -				
8" x 10"	20 Ton	2,050.00	11,000.00	7,500.00
8" x 10"	30 Ton	2,675.00	12,000.00	7,600.00
10" x 10"	20 Ton	2,850.00	12,500.00	8,100.00
10" x 10"	30 Ton	3,250.00	14,100.00	8,200.00
Add to Above for Printer and Dial			3,000.00	

.3 CRANE SCALES -

	1 Ton	157.00	2,850.00	-
	5 Ton	305.00	3,100.00	-

1314.0 SIGNS (Special Exterior) (L&M) (Sheet Metal Workers)

1315.0 SKYDOMES (L&M) (Carpenters)

EXTRUDED ALUMINUM FRAME, WIRE GLASS EXTERIOR AND FIBERGLASS SUB CEILING

	UNIT	COST
12' x 18' Ridge Type	Each	8,800.00
	or SqFt	40.75
18' x 18' Pyramid Type	Each	17,000.00
	or SqFt	52.50
36' Diameter Geometric Domes	Each	69,800.00
	or SqFt	67.20
18' Diameter Revolving Domes	Each	18,000.00
	or SqFt	64.60

1316.0 SOUND INSULATED ROOMS (Including Vibration Control) (M) (Carpenters)

.1 ANECHOIC ROOMS (ISOLATORS, PANELS AND WEDGES)	Room	131,200.00
Example: Room 30' x 30'	or SqFt	145.80
.2 AUDIOMETRIC ROOM (PREFABRICATED PANELS)		
Example: Research Room - 5' x 8'	Room	11,600.00
	or SqFt	290.00
Example: Medical Practice Room - 5' x 8'	Room	26,200.00
	or SqFt	655.00

1318.0 SWIMMING POOLS

.1 POOLS

	UNIT	COST
<u>Deep Pools</u> - 3 to 9 feet		
(Incl. Filters, Skimmers, Lights, Clng. Equip.)		
Rectangular or L Shape:		
Residential - 1,000 SqFt	Each	41,500.00
	or SqFt	41.50
Motels and Apartments - 2,000 SqFt	Each	96,600.00
	or SqFt	48.30
Municipal - 5,000 SqFt	Each	278,000.00
	or SqFt	55.60
Add for Diving Stand - Steel - 1 Meter	Each	3,500.00
Steel - 3 Meter	Each	5,200.00
Add for Diving Boards - 12' Fiberglass	Each	910.00
16' Aluminum	Each	1,650.00
Add for Slides - 6' Fiberglass	Each	870.00
12' Fiberglass	Each	1,350.00
Add for Ladders - 10'	Each	335.00
Add for Lifeguard Chairs	Each	1,130.00
Add for Covers	Each	810.00
<u>Shallow Pools</u> - 3 to 5 feet -		
Gunite with Plaster Finish		
Residential - 1,000 SqFt	Each	31,500.00
	or SqFt	31.50
Motels and Apartments - 2,000 SqFt	Each	80,800.00
	or SqFt	40.40
Deduct for Vinyl Lined Pools	SqFt	9.70
Add for Free Form Pools	Each	5%
Add for Heaters	Each	4,200.00
Add for Automatic Cleaners	Each	4,000.00
<u>Wading Pools</u> - 12' x 12' or 9' Round	Each	5,100.00
	or SqFt	35.40
.2 DECKS - CONCRETE - Average 10' around Pool	Each	4,000.00
and Average 1,000 SqFt	or SqFt	4.00
Add for Tile	SqFt	8.40
Add for Brick	SqFt	7.20
.3 FENCES - 4' ALUMINUM OR GALVANIZED	LnFt	1,190.00

1319.0 TANKS

	UNIT	COST
.1 STEEL - GROUND LEVEL (NO FOUNDATION)		
150,000 Gals	Each	150,000.00
500,000 Gals	Each	253,000.00
1,000,000 Gals	Each	420,000.00
2,000,000 Gals	Each	660,000.00
.2 STEEL - ELEVATED		
100,000 Gals	Each	325,000.00
250,000 Gals	Each	435,500.00
500,000 Gals	Each	705,000.00
1,000,000 Gals	Each	1,550,000.00

1320.0 WOOD DECKS - WITH RAILS

	UNIT	COST
CEDAR	SqFt	16.20
TREATED FIR	SqFt	14.40
Add for Seats	LnFt	4.00
Add for Stairs	Riser	55.00

1401.0 ELEVATORS (L&M) (Elevator Constructors & Ironworkers)

		Speed to F.P.M.	2,000#	3,000#	4,000#	10,000#
.1	ELECTRIC (CABLE/ TRACTION)		Capacity and Cost			
.11	Passenger - Std Specifications					
	Geared A.C. Rheostatic:					
	Selective Collective Operation					
	3 Openings	125	72,000	-	-	-
	4 Openings	125	75,200	-	-	-
	5 Openings	125	78,400	-	-	-
	Add per Floor	-	5,300	-	-	-
	Geared Variable Voltage:					
	Selective Collective Operation					
	8 Openings	250	130,400	137,500	140,000	-
		350	134,600	138,500	143,000	-
	10 Openings	250	133,500	137,500	145,000	-
		350	137,500	135,500	148,000	-
	Add per Floor	-	6,900	7,400	7,900	-
	Group Supervisory					
	8 Openings	250	-	147,000	149,000	-
		350	-	155,000	154,000	-
	10 Openings	250	-	147,000	152,000	-
		350	-	154,000	155,000	-
	Add per Floor	-	-	7,000	7,500	-
	Gearless Variable Voltage:					
	Group Supervisory					
	8 Openings	500	-	179,100	175,000	-
		800	-	192,900	189,000	-
		1000	-	202,000	204,000	-
	10 Openings	500	-	181,300	163,200	-
		800	-	198,200	203,500	-
		1000	-	217,300	205,000	-
	Add per Floor	-	-	7,200	7,400	-
	Add per Fl. of Express Zone	-	6,700	7,100	-	-
.12	Freight					
	3 Openings	100	-	-	95,400	100,000
	4 Openings	100	-	-	100,000	105,000
	Add per Floor	-	-	-	22,500	12,800
	Add per Power Operated Door		-	-	-	7,400
.2	OIL HYDRAULIC					
.21	Passenger - 2 Openings	100	52,000	-	-	-
	3 Openings	100	56,000	-	-	-
	4 Openings	100	63,000	-	-	-
	5 Openings	100	70,000	-	-	-
	6 Openings	100	80,000	-	-	-
	Deduct for Limited Use Elevator, non-Commercial 8,000# - $5,000					
.22	Freight - 2 Openings	100	-	-	60,000	74,000
	3 Openings	100	-	-	71,000	78,000
	4 Openings	100	-	-	73,000	81,000
	Add per Power Operated Door - $11,000					

.3 RESIDENTIAL (WINDING DRUM)

2 Openings - Speed to 36 F.P.M., 700# Capacity - $21,000

Speed to 30 F.P.M., 410# Capacity - $16,500

Add for 3 Openings - $3,800

Add for ADA Updating -$13,500

Add for Updating Generators, Controllers & Operating Equipment -$60,000

Add to 1401.1 and 1401.2 Above for Custom Specifications - 10%

Add to 1401.2 for Shaft Work, Electric Power & Mechanical - approx. $60,000

Add for Demolition – Electric (5 openings) $18,000

Add for Demolition – Hydraulic (5 openings) $10,800

		Speed F.P.M.	UNIT	CAPACITY	COST
1402.0	**MOVING STAIRS & WALKS (L&M) (Elevator Constructors & Ironworkers)**				
.1	MOVING STAIRS (Escalators)				
	16' Floor Height				
	Metal Balustrade 32" Wide				125,000
	48"				136,000
	Glass Balustrade 32" Wide				135,000
	48"				148,000
	Add or Deduct per Foot Floor Height				3,200
	Add for Over 20 Feet Floor Height				25%
.2	MOVING WALKS - Belt Type 32"		LnFt		810
	48"		LnFt		900
	Pallet Type 32"		LnFt		990
	48"		LnFt		1,000
1403.0	**DUMBWAITERS (L&M) (Elevator Constructors)**				
.1	ELECTRIC				
	Drum Type				
	2 Openings - 1 Stop	25 to 75		50# to 500#	15,500
	3 Openings - 2 Stops	25 to 75		50# to 500#	18,900
	Add per Floor				4,100
	Traction Type				
	3 Openings - 2 Stops	50 to 300		50# to 500#	20,400
	4 Openings - 3 Stops	50 to 300		50# to 500#	23,500
	5 Openings - 4 Stops	50 to 300		50# to 500#	27,800
	6 Openings - 5 Stops	50 to 300		50# to 500#	30,500
	Add per Floor				4,200
.2	HAND OPERATED				
	2 Openings - 1 Stop				6,750
	3 Openings - 2 Stops				7,500
	Add per Floor				1,020
	Add for all Stainless Steel				25%
1404.0	**LIFTS (L&M) (Ironworkers & Elevator Constructors)**				
.1	WHEELCHAIR LIFTS - 42"		Each	1,050#	18,900
	96"		Each	1,400#	22,700
.2	STAIR LIFTS (CLIMBERS)		Each		6,500
.3	VEHICLE - SEMI HYDRAULIC LIFTS - 1 Post		Each	8,000#	7,800
	2 Post		Each	10,000#	12,400
	2 Post		Each	15,000#	19,000
	Add for Fully Hydraulic - 2 Post		Each	10,000#	4,350
.4	MATERIAL HANDLING LIFTS		Each		-
.5	STAGE LIFTS - 12' x 30'		Each		91,700
	Add for Permits, Inspections & Elec Hookups				1,600

See 1116 for Loading Dock Lift Equipment

		UNIT	COST
1405.0	**PNEUMATIC TUBE SYSTEMS (L&M) (Sheet Metal Workers)**		
	3" ROUND - Single Tube	LnFt	32
	Twin Tube	LnFt	42
	4" ROUND - Single Tube	LnFt	33
	Twin Tube	LnFt	44
	4" x 7" OVAL - Single Tube	LnFt	40
	Twin Tube	LnFt	56
1406.0	**MATERIAL HANDLING (L&M) (Ironworkers & Millwrights)**		-
1407.0	**TURNTABLES (L&M) (Ironworkers & Millwrights)**		-
1408.0	**POWERED SCAFFOLDING (L&M) (Ironworkers)**		-

Costs of mechanical work are priced as total contractor's costs with an overhead and fee of 20% included. Not included are scaffold, hoisting, temporary heat, enclosures and bonding. Included are 35% taxes and insurance on labor, equipment, tools and 5% sales tax.

1501.0 BASE MATERIALS AND METHODS	UNIT	COST
.1 PIPE AND PIPE FITTINGS (Including Fittings)		
Steel - Black - Threaded 3/4"	LnFt	3.35
1"	LnFt	3.85
1-1/2"	LnFt	5.60
2"	LnFt	10.60
2-1/2"	LnFt	15.20
3"	LnFt	20.70
4"	LnFt	25.50
5"	LnFt	29.00
6"	LnFt	38.00
8"	LnFt	55.00
Steel - Galvanized Schedule 40 3/4"	LnFt	7.00
1"	LnFt	9.00
1-1/4"	LnFt	10.20
1-1/2"	LnFt	12.00
2"	LnFt	15.50
3"	LnFt	25.70
4"	LnFt	34.20
5"	LnFt	42.00
6"	LnFt	57.00
8"	LnFt	62.00
Deduct for Black Iron	LnFt	5%
Deduct for Underground	LnFt	10%
Cast Iron - Hub and Heavy Duty 2"	LnFt	16.20
3"	LnFt	16.80
4"	LnFt	23.40
5"	LnFt	28.20
6"	LnFt	38.20
8"	LnFt	45.00
Add for Line Pipe	LnFt	4.60
Deduct for Hubless Type	LnFt	10%
Deduct for Standard	LnFt	15%
P.V.C. - Schedule 40 1/2"	LnFt	4.90
3/4"	LnFt	5.40
1"	LnFt	5.95
1-1/4"	LnFt	6.50
1-1/2"	LnFt	7.15
2"	LnFt	7.80
3"	LnFt	8.90
4"	LnFt	15.20
6"	LnFt	26.25
Deduct for Underground	LnFt	10%
P.V.C. - Type L 3/8"	LnFt	3.60
1/2"	LnFt	4.80
3/4"	LnFt	5.60
1"	LnFt	6.50
1-1/4"	LnFt	7.20
1-1/2"	LnFt	8.00
2"	LnFt	9.00
3"	LnFt	10.50
4"	LnFt	15.60
6"	LnFt	30.00
Add for Type K (Underground)	LnFt	20%
Deduct for Type M (Residential)	LnFt	5%

1501.0 BASE MATERIALS AND METHODS (Cont'd...)

.1 PIPE AND PIPE FITTINGS (Cont'd...)	UNIT	COST
Brass - Threadless 1/2"	LnFt	9.30
3/4"	LnFt	10.50
1"	LnFt	13.00
1-1/4"	LnFt	16.00
1-1/2"	LnFt	19.00
2"	LnFt	23.50
3"	LnFt	33.00
4"	LnFt	43.00
Add for Threaded	LnFt	30%
Stainless Steel Schedule 40 1/4"	LnFt	8.50
1/2"	LnFt	12.70
3/4"	LnFt	16.20
1"	LnFt	19.00
2"	LnFt	26.00
4"	LnFt	64.00
Add for Type 316	LnFt	20%
Fibre Glass - 1"	LnFt	16.90
1-1/2"	LnFt	19.30
2"	LnFt	22.00
3"	LnFt	25.50
4"	LnFt	29.00
5"	LnFt	37.00
6"	LnFt	55.00
8"	LnFt	65.00

Utility Piping -

	UNIT	Reinforced Concrete	Vitrified Clay	Bituminous Coated Corrugated	Smooth Paved Corrugated	Ductile Iron
6"	LnFt	$7.10	$6.70	$9.40	$9.90	$15.50
8"	LnFt	7.60	9.30	9.70	10.70	21.00
1"	LnFt	9.60	12.20	11.40	12.60	24.50
12"	LnFt	13.20	14.40	14.60	15.50	37.00
15"	LnFt	15.30	21.70	16.80	18.00	46.00
18"	LnFt	20.70	33.00	20.40	31.00	60.00
21"	LnFt	25.40	38.70	24.20	27.00	73.00
24"	LnFt	28.80	55.50	26.85	31.80	88.00
27"	LnFt	39.00	77.00	32.60	37.00	-
30"	LnFt	48.00	101.00	33.80	43.00	-
36"	LnFt	58.00	140.00	37.50	54.00	-
42"	LnFt	75.00	-	43.90	62.00	-
48"	LnFt	97.00	-	59.90	70.00	-
54"	LnFt	112.00	-	79.20	94.00	-
60"	LnFt	132.00	-	96.30	118.00	-

Add for Trenching and Backfill - See Division 0202.0
See 0205 for Site Drainage Items

1501.0 BASE MATERIALS AND METHODS, Cont'd...

		UNIT	COST
.2	PIPING SPECIALTIES		
.21	Expansion Joints - Copper 1/2"	Each	$66.00
	3/4"	Each	100.00
	1"	Each	155.00
	1-1/2"	Each	330.00
	2"	Each	500.00
	3"	Each	940.00
	4"	Each	1020.00
.22	Vacuum Breakers - 1/2"	Each	40.00
	3/4"	Each	46.00
	1"	Each	62.00
	1-1/2"	Each	86.00

		UNIT	IRON BODY	BRONZE BODY
.23	Strainers - Screwed Ends - 1/2"	Each	33.00	43.50
	3/4"	Each	38.00	50.00
	1"	Each	45.00	67.00
	1-1/2"	Each	62.00	105.00
	2"	Each	100.00	175.00
	3"	Each	175.00	335.00
	4"	Each	470.00	1,280.00
	Strainers - Flanged Ends - 2"	Each	160.00	445.00
	3"	Each	280:00	925.00
	4"	Each	460.00	1,600.00
	5"	Each	720.00	2,200.00
	6"	Each	960.00	3,050.00

.3 MECHANICAL - SUPPORTING ANCHORS & SEALS

.4 VALVES

Bronze	UNIT	150 psi Gate	150 psi Check	150 psi Globe	Relief	300 psi to 75 Reducing
Body - 1/2"	Each	$30.00	$27.00	$36.00	$54.00	$73.00
3/4"	Each	37.00	30.00	47.00	63.00	87.00
1"	Each	44.00	36.00	64.00	92.00	127.00
1-1/2"	Each	71.00	58.00	99.00	300.00	285.00
2"	Each	92.00	88.00	143.00	335.00	425.00
2-1/2"	Each	165.00	520.00	250.00	-	-
3"	Each	225.00	192.00	370.00	-	-

Iron	UNIT	125 psi Gate	125 psi Check	125 psi Globe
Body - 2"	Each	$265.00	$210.00	$380.00
2-1/2"	Each	275.00	215.00	435.00
3"	Each	310.00	220.00	510.00
4"	Each	465.00	400.00	680.00
6"	Each	650.00	425.00	870.00
		Gate	Check	Globe
Steel - 2"	Each	$890.00	$870.00	$1,120.00
2-1/2"	Each	1,390.00	1,150.00	1,620.00
3"	Each	1,400.00	1,150.00	1,730.00
4"	Each	1,670.00	1,620.00	2,420.00

		Ball	Foot
Plastic - 1/2"	Each	$24.00	$57.00
3/4"	Each	29.00	63.00
1"	Each	33.50	83.00
1-1/2"	Each	58.00	133.00
2"	Each	66.00	160.00

1501.0	BASE MATERIALS AND METHODS, Cont'd...	UNIT	COST
.5	PUMPS		
	Condensate Return - Simplex	Each	$ 1,850.00
	Duplex	Each	4,950.00
	Utility Water Pump -		
	Multi - Two Stage 3" & 4" 75 HP	Each	18,500.00
	Four Stage 3" & 4" 150 HP	Each	34,500.00
	Single - End Suction 1" x 2" 3 HP	Each	4,850.00
	End Suction 2" x 3" 15 HP	Each	6,150.00
	Condenser Water Pump, Pressure - 100 GPM	Each	16,500.00
	300 GPM	Each	22,000.00
	1200 GPM	Each	49,000.00
	2000 GPM	Each	59,500.00
	Submersible Pump 2"	Each	685.00
	Pedestal Pump	Each	210.00
.6	VIBRATION ISOLATION 5" long	Each	345.00
	6" long	Each	670.00
.7	METERS AND GAGES		
	Disc Type - 3/4"	Each	175.00
	1"	Each	215.00
	2"	Each	600.00
	Compound Type - 3"	Each	2,100.00
	4"	Each	3,260.00
.8	TANKS		
	Fuel Storage - Steel - 1,000 Gallons	Each	2,050.00
	2,000	Each	2,850.00
	5,000	Each	6,500.00
	10,000	Each	9,300.00
	Fuel Storage - Fiberglass - 2,000 Gallons	Each	3,300.00
	5,000	Each	5,750.00
	10,000	Each	9,600.00
	Hot Water Storage - 80 Gallons	Each	395.00
	140	Each	600.00
	200	Each	830.00
	400	Each	1,325.00
	600	Each	2,000.00
	Liquid Expansion - 30 Gallons	Each	350.00
	60	Each	560.00
	100	Each	775.00

1502.0 MECHANICAL INSULATION

.1	INSULATION FOR PIPING	Unit	3/8" Foam	1" Wall Fiberglass	2" Wall Fiberglass
	Interior - Pipe 1/2"	LnFt	3.60	4.45	6.25
	3/4"	LnFt	3.70	4.70	6.65
	1"	LnFt	4.05	4.90	6.75
	1-1/4"	LnFt	4.30	5.30	7.00
	1-1/2"	LnFt	4.55	5.70	7.40
	2"	LnFt	5.00	6.00	8.20
	2-1/2"	LnFt	5.45	6.40	8.95
	3"	LnFt	5.85	6.90	10.40
	6"	LnFt	8.30	8.70	14.30
	8"	LnFt	10.10	12.60	15.80
	Exterior - with Aluminum Jacket - Add 2.00/ LnFt				

.2	INSULATION FOR DUCT WORK	Unit	1"	1-1/2"	2"
	Rigid with Vapor Barrier	SqFt	2.30	3.25	2.65
	Blanket	SqFt	4.80	3.55	2.90
.3	INSULATION FOR BOILERS				
	Fiberglass - 2"	SqFt	-	-	11.10
	Calcium Silicate - 2" with C.F.	SqFt	-	-	16.80

		UNIT	COST
1503.0	**FIRE PROTECTION EQUIPMENT**		
.1	AUTOMATIC SPRINKLER EQUIPMENT - Wet System	SqFt	2.20
		(or) Head	210.00
	Dry System	SqFt	2.50
		(or) Head	225.00
	Add for Pendant Type Sprinklers	SqFt	.60
	Add for Deluge System	SqFt	2.10
.2	CARBON DIOXIDE EQUIPMENT - Cylinder	Each	1,000.00
	Detector & Nozzle	Each	335.00
.3	STANDPIPE AND FIRE HOSE EQUIPMENT		
	Standpipe - Exposed 6"	Each	625.00
	Concealed 6"	Each	800.00
	Cabinets - 30" x 30" Aluminum	Each	320.00
	Hose and Nozzle - 1 1/2"	LnFt	2.90
	2 1/2"	LnFt	3.50
1504.0	**PLUMBING**		
.1	EQUIPMENT	-	-
.2	PACKAGE WASTE, VENT OR WATER PIPING	Each	470.00
.3	DOMESTIC WATER SOFTENER	Each	880.00
.4	PLUMBING FIXTURES (Drain, Waste and Vent Not Included - See 1504.5)		
	Porcelain Enameled (unless otherwise noted) with accessories		
	Bathtub - 5' - Cast Iron - Enameled	Each	1,150.00
	Steel - Enameled	Each	580.00
	Fiberglass	Each	600.00
	5' x 5' - Cast Iron - Enameled	Each	1,500.00
	Bathtub/ Whirlpool - Fiberglass 6'	Each	4,100.00
	Bidet	Each	1,000.00
	Drinking Fountain - Enameled Wall Hung	Each	610.00
	SS Recessed	Each	740.00
	Garbage Disposal - 1/2 Horsepower	Each	170.00
	Lavatories - 19" x 17"- Oval	Each	305.00
	20" x 18" - Rectangular	Each	350.00
	27" x 23" - Oval Pedestal	Each	755.00
	18" - SS Bowl	Each	385.00
	Add for Handicap Lavatory	Each	190.00
	Shower & Tub - Combined - Fiberglass	Each	1,400.00
	Shower/Stall - Fiberglass - 32" x 32"	Each	830.00
	Fiberglass - 36" x 36"	Each	1,100.00
	Sink, Kitchen - 24" x 21" - Single	Each	535.00
	24" x 21" - SS Single	Each	373.00
	33" x 22" - Double	Each	590.00
	32" x 21" - SS Double	Each	450.00
	Sink, Laundry - Enameled	Each	590.00
	Plastic	Each	385.00
	Sink, Service - 24" x 21"	Each	725.00
	Urinals - Wall Hung	Each	645.00
	Stall or Trough 4'	Each	860.00
	Wash Fountain - Circular - Precast Terrazzo	Each	2,100.00
	Semi-circular - Precast Terrazzo	Each	2,100.00
	Water Closets - Floor Mounted - Tank Type	Each	470.00
	One Piece	Each	635.00
	Wall Mounted	Each	655.00
	Add for Colored Fixtures	Each	25%
	Add for Handicap Water Closet	Each	105.00
	See 1106 for Appliances		
.5	RESIDENTIAL PLUMBING		
	Average Domestic Fixture, Rough In and Trim	Each	525.00
	Average Cost of Drain, Waste, and Vent	Each	900.00
	Floor Drains - 4"	Each	475.00
.6	POOL EQUIPMENT - See 1318.0	-	-
.7	FOUNTAIN PIPING	-	-

		UNIT	COST
1505.0	**HEAT GENERATION**		
.1	FUEL HANDLING EQUIPMENT		-
.2	ASH REMOVAL SYSTEM - Electric	Each	8,900.00
	Hand	Each	5,750.00
.3	LINED BREECHINGS		-
.4	BOILERS		
	Package Type - 1,000 Lbs per Hour	Each	27,500.00
	Steam & Hot Water - 2,000 Lbs per Hour	Each	34,000.00
	5,000 Lbs per Hour	Each	57,000.00
	10,000 Lbs per Hour	Each	63,500.00
	15,000 Lbs per Hour	Each	88,000.00
	20,000 Lbs per Hour	Each	95,000.00
	25,000 Lbs per Hour	Each	130,000.00
	Cast Iron - 100 MBH	Each	2,500.00
	Hot Water or Steam - 200 MBH	Each	4,300.00
	300 MBH	Each	5,450.00
	600 MBH	Each	9,150.00
	1,000 MBH	Each	15,000.00
	Hot Water Heaters - 20 GPM	Each	640.00
	30 GPM	Each	840.00
	40 GPM	Each	920.00
	50 GPM	Each	1,250.00
1506.0	**REFRIGERATION**		
.1	REFRIGERANT COMPRESSORS		-
.2	CONDENSERS - Air Cooled - 10 Ton	Each	14,700.00
	20 Ton	Each	17,700.00
	30 Ton	Each	25,500.00
	40 Ton	Each	30,800.00
	60 Ton	Each	43,800.00
	80 Ton	Each	52,200.00
	Add for Water Cooled	Each	10%
.3	CHILLERS - Hermetic Centrifugal - 100 Ton	Each	57,300.00
	150 Ton	Each	59,800.00
	300 Ton	Each	83,200.00
	Add for Absorption Type	Each	15%
.4	COOLING TOWERS - 100 Ton	Each	7,900.00
	150 Ton	Each	14,700.00
	300 Ton	Each	39,600.00
1507.0	**HEAT TRANSFERS**		
.1	HOT WATER SPECIALTIES		-
.2	CONDENSATE PUMPS AND RECEIVER SETS		
	Outdoor - Air Cooled - 30,000 BTU	Each	2,000.00
	50,000 BTU	Each	2,700.00
	Indoor - 30,000 BTU	Each	1,020.00
	50,000 BTU	Each	1,400.00
.3	HEAT EXCHANGERS - 100 GPM	Each	1,200.00
	200 GPM	Each	1,290.00
	300 GPM	Each	2,800.00
	600 GPM	Each	5,700.00
	1,000 GPM	Each	7,600.00
.4	TERMINAL UNITS		
	Base Boards - Cast Iron Radiant - 7 1/4"	LnFt	23.50
	9 3/4"	LnFt	25.00
	Tube with Cover - 7 1/4" - 3/4"	LnFt	13.00
	7 1/4" - 1"	LnFt	14.10
	Tube with Cover - 9 3/4" - 3/4"	LnFt	16.20
	9 3/4" - 1"	LnFt	19.90
	14" Fin Tube - Steel	LnFt	24.40
	Copper	LnFt	29.70
	Convectors	Each	290.00
	Radiators	SqFt	8.60

		UNIT	COST
1507.0	**HEAT TRANSFERS, Cont'd...**		
.5	COILS	-	-
.6	UNIT HEATERS - 100,000 BTU	Each	730.00
	150,000 BTU	Each	1,000.00
	200,000 BTU	Each	1,150.00
.7	PACKAGED HEATING AND COOLING	-	2,300.00
.8	STEAM SPECIALTIES	-	520.00
1508.0	**AIR HANDLING AND DISTRIBUTION**		
.1	FURNACE - Duct - 100 MBH	Each	1,275.00
	200 MBH	Each	1,900.00
	300 MBH	Each	3,250.00
.2	FANS - 1,000 CFM	Each	140.00
	5,000 CFM	Each	590.00
	10,000 CFM	Each	980.00
.3	DUCT WORK - Galvanized - 26 Ga	Lb	3.25
	24 Ga	Lb	3.45
	22 Ga	Lb	3.90
	20 Ga	Lb	4.25
.4	DUCT ACCESSORIES		
	Constant Volume Duct	Each	580.00
	Reheat Constant Volume Duct	Each	715.00
	Diffusers & Registers - Sidewalk	SqFt	30.00
	Ceiling	SqFt	50.00
.5	RESIDENTIAL HEATING AND AIR CONDITIONING		
	Furnace	Each	1,420.00
	Central Air Conditioning	Each	670.00
	Humidifier	Each	330.00
	Air Cleaner	Each	535.00
	Thermostat	Each	140.00
	Attic Fan	Each	320.00
	Ventilating Damper	Each	120.00
	6" Flue	Each	365.00
.6	COMMERCIAL AIR CONDITIONING		
	Average	Ton	2,450.00
	High	Ton	3,800.00
	Low	Ton	1,200.00
1509.0	**SEWAGE DRAINAGE**		
.1	SEWAGE EJECTORS 100 GPM	Each	7,600.00
.2	GREASE INTERCEPTORS - 200 GPM	Each	8,800.00
	300 GPM	Each	16,700.00
.3	LIFT STATIONS - STEEL, CONCRETE OR FIBERGLASS		
	WITH GENERATOR - 100 GPM	Each	76,000.00
	200 GPM	Each	130,000.00
	500 GPM	Each	190,000.00
.4	SEPTIC TANKS AND DRAIN FIELD		
	Concrete Tanks with Excav. - 1,000 Gallons	Each	1,520.00
	2,000	Each	3,300.00
	5,000	Each	7,000.00
	10,000	Each	13,200.00
	Drain Fields	LnFt	5.00
.5	SEWAGE TREATMENT EQUIPMENT - STEEL OR CONCRETE		
	5,000 Gallons	GPD	9.70
	50,000	GPD	3.50
	100,000	GPD	3.30
.6	AERATION EQUIPMENT - STEEL - 1,000 Gallons	Each	18,000.00
	CAPACITY PER DAY 2,000	Each	27,000.00
	5,000	Each	36,500.00
	10,000	Each	44,000.00
.7	SLUDGE DIGESTION		-
1510.0	**CONTROLS AND INSTRUMENTATION - No cost for items listed on Index**		-
1511.0	**TESTING, ADJUSTING AND BALANCING**		

Costs of Electrical Work are priced as total contractors' costs with an overhead and fee of 20% included.

1601.0 BASIC MATERIALS AND METHODS

.1 CONDUITS (Incl Fittings/Supports)

		UNIT	1/2"	3/4"	1-1/2"	2"	3"	4"	5"	6"
			COST							
.11	Rigid									
	Galvanized - Rigid	LnFt	5.50	6.80	10.60	14.70	25.90	35.90	60.50	84.00
	EMT	LnFt	3.60	4.60	6.35	8.00	14.15	20.10	-	-
	PVC	LnFt	3.20	3.45	4.90	5.90	11.55	15.45	-	-
	Add for Running Exposed									
.12	Flexible									
	Greenfield	LnFt	3.30	3.75	7.20	8.70	17.00	-	-	-
	Sealtite	LnFt	4.85	6.00	12.65	16.00	35.20	45.00	-	-
.13	Ducts									
	Fibre	LnFt	-	-	-	-	4.30	4.60	5.25	5.85
	Transite	LnFt	-	-	-	-	4.40	5.10	5.70	6.30

Cost for Conduits includes Labor, Material, Equipment and Fees and Installed in Wood or Steel Construction.

.2 BOXES, FITTINGS AND SUPPORTS (In Above)

		UNIT	COST Copper T.H.W.
.3	WIRE AND CABLE		
	Wire - 3 - #14 - 600 Volt	LnFt	1.70
	3 - #12 - 600 Volt	LnFt	2.10
	3 - #10 - 600 Volt	LnFt	3.00
	3 - # 8 - 600 Volt	LnFt	3.60
	3 - # 4 - 600 Volt	LnFt	5.20
	3 - # 2 - 600 Volt	LnFt	6.90
	3 - 1/0	LnFt	9.50
	3 - 2/0	LnFt	11.20
	3 - 3/0	LnFt	12.50
	3 - 4/0	LnFt	14.80
	3 - 250 - MCM	LnFt	17.50
	3 - 300 - MCM	LnFt	19.80
	3 - 350 - MCM	LnFt	22.00
	3 - 400 - MCM	LnFt	25.00
	3 - 500 - MCM	LnFt	27.40

	UNIT	ARMORED COPPER 2 Wire	3 Wire	NON-METALLIC 2 Wire	3 Wire
Cable - #14 bx - 600 Volt	LnFt	3.40	3.80	1.60	2.90
#12 bx - 600 Volt	LnFt	3.60	4.10	1.75	3.00
#10 bx - 600 Volt	LnFt	4.20	5.00	2.20	3.30
# 8 bx - 600 Volt	LnFt	-	6.25	-	3.60
# 6 bx - 600 Volt	LnFt	-	7.00	-	4.00
# 4 bx - 600 Volt	LnFt	-	8.00	-	4.80
Add for Work over 10'					10%
Add for Work in Masonry Construction					10%
Add for Work in Concrete Construction					20%

.4	SWITCH GEAR	
	Metal Clad 5 KV	22,000.00
	15 KV	26,500.00
	Add for Outdoor Type	4,400.00

1601.0 BASIC MATERIALS AND METHODS, Cont'd...

		UNIT	COST
.5	SWITCHES AND RECEPTACLES		
	Switches		
	Standard Toggle - 1 Pole	Each	70.00
	4 W	Each	125.00
	4 WL	Each	140.00
	Safety - 30 Amp	Each	180.00
	60 Amp	Each	225.00
	100 Amp	Each	360.00
	200 Amp	Each	550.00
	Three-way -	Each	110.00
	Add for Dimmer - 600 Watt	Each	80.00
	Add for Dimmer - 1500 Watt	Each	170.00
	Receptacles		
	Single	Each	70.00
	Duplex - GFI	Each	90.00
	Add for Dedicated - Separate Circuit	Each	150.00
	Add for Dedicated - Ground	Each	20.00
	Add for Dedicated - Fourplex	Each	70.00
	Clock, Lamps, etc.	Each	72.00
	Add for Weatherproof	Each	18.00
	Plates - Stainless Steel	Each	10.00
	Brass	Each	15.00
.6	TRENCH DUCT AND UNDERFLOOR DUCT (Including Fittings)		
	Underfloor Duct - Average		
	Single Cell - 1 1/4" x 3 1/8"	LnFt	18.00
	1 1/2" x 6 1/2"	LnFt	23.00
	Double Cell - 1 1/4" x 3 1/8"	LnFt	26.00
	1 1/2" x 6 1/2"	LnFt	45.00
	Wireway - Cover Type 2 1/2" x 2 1/2" x 1"	Each	19.00
	2 1/2" x 2 1/2" x 2"	Each	26.00
	2 1/2" x 2 1/2" x 3"	Each	31.00
	4" x 4" x 1"	Each	24.00
	4" x 4" x 2"	Each	30.00
	4" x 4" x 3"	Each	38.00
.7	MOTOR WIRING (Including Controls)		
	10 H.P.	Each	3,900.00
	20 H.P.	Each	4,000.00
	30 H.P.	Each	4,500.00
	40 H.P.	Each	4,800.00
	50 H.P.	Each	5,700.00
	100 H.P.	Each	7,400.00
	200 H.P.	Each	11,500.00
.8	MOTOR STARTERS (600 Volt)		
	Magnetic - 10 H.P.	Each	480.00
	Non-Reversing - 20 H.P.	Each	560.00
	30 H.P.	Each	840.00
	40 H.P.	Each	950.00
	50 H.P.	Each	1,400.00
	100 H.P.	Each	2,200.00
	200 H.P.	Each	4,400.00
	Add for Reversing Type	Each	80%
	Add for Explosion Proof	Each	40%
	Add for Motor Circuit Protectors	Each	25%
	Add for Fused Switches	Each	10%
	Manual - 1 H.P.	Each	120.00
	3 H.P.	Each	170.00
	7 1/2 H.P.	Each	200.00

1601.0 BASIC MATERIALS AND METHODS, Cont'd...

		UNIT	COST
.9	CABLE TRAY (Aluminum or Galvanized)		
	Ladder or Punched - 6" wide	LnFt	27.00
	12" wide	LnFt	31.00
	18" wide	LnFt	33.00
	24" wide	LnFt	35.00

1602.0 POWER GENERATION

	UNIT	COST
Emergency Generators (3-Phase, 4-Wire) (120/208 Volt)		
15 k.w.	Each	13,000.00
30 k.w.	Each	26,000.00
50 k.w.	Each	32,000.00
75 k.w.	Each	35,000.00
100 k.w.	Each	45,000.00
300 k.w.	Each	75,000.00

1603.0 HIGH VOLTAGE DISTRIBUTION - Power Companies

1604.0 SERVICE AND DISTRIBUTION

		UNIT	COST
.1	METERING		
.2	GROUNDING (Including Rods and Accessories)		
	1/2" x 10'	Each	95.00
	5/8" x 10'	Each	105.00
	3/4" x 10'	Each	120.00
.3	SERVICE DISCONNECTS - 100 amp	Each	300.00
	200 amp	Each	550.00
	400 amp	Each	900.00
	600 amp	Each	1,750.00
.4	DISTRIBUTION SWITCHBOARDS (240 Volt) - 400 amp	Each	3,500.00
	800 amp	Each	3,600.00
	1600 amp	Each	5,200.00
.5	LIGHT AND POWER PANELS (3-Phase, 4-Wire) (120/208 Volt)		
	12 circuit	Each	720.00
	16 circuit	Each	860.00
	20 circuit	Each	1,000.00
	24 circuit	Each	1,200.00
	30 circuit	Each	1,400.00
	36 circuit	Each	1,550.00
	Add for Circuit Breakers	Each	230.00
.6	TRANSFORMERS		
	Current - 400 amp	Each	3,100.00
	600 amp	Each	3,400.00
	800 amp	Each	3,800.00
	1000 amp	Each	4,200.00

Light and Power	UNIT	SINGLE-PHASE 120/ 240 Volt	3-PHASE 120/ 208 Volt
Dry Type - 3 kva	Each	360.00	530.00
5 kva	Each	520.00	750.00
10 kva	Each	730.00	970.00
15 kva	Each	1,020.00	1,350.00
30 kva	Each		1,800.00
45 kva	Each		2,300.00
75 kva	Each		2,500.00
Oil Filled - 150 kva	Each	-	5,600.00
300 kva	Each	-	10,600.00
500 kva	Each	-	13,800.00

1605.0 LIGHTING

.1 LIGHT FIXTURES	UNIT	COST
Fluorescent		
Surface		
Industrial - 48" - 1 lamp R.S.	Each	105.00
H.O.	Each	130.00
2 lamp R.S.	Each	125.00
H.O.	Each	160.00
4 lamp R.S.	Each	140.00
H.O.	Each	240.00
96" - 1 lamp H.O.	Each	140.00
R.S.	Each	240.00
2 lamp H.O.	Each	120.00
H.O.	Each	245.00
Commercial - 12" x 48" - 2 lamp R.S.	Each	140.00
24" x 24" - 4 lamp R.S.	Each	185.00
24" x 48" - 4 lamp R.S.	Each	245.00
48" x 48" - 4 lamp R.S.	Each	370.00
12" x 48" - 2 lamp R.S.	Each	175.00
Recessed - 24" x 24" - 4 lamp R.S.	Each	190.00
24" x 48" - 4 lamp R.S.	Each	220.00
48" x 48" - 4 lamp R.S.	Each	380.00
Add for Electronic Ballasts	Each	22.00
Incandescent - 100 watt	Each	90.00
200 watt	Each	105.00
300 Watt	Each	120.00
Quartz - 500 watt	Each	165.00
Sodium - Low Pressure - 35 watt	Each	290.00
55 watt	Each	430.00
Explosion Proof		
Fluorescent- 48" - 2 lamp R.S.	Each	1,030.00
Incandescent - 48" - 200 watt	Each	330.00
Mercury Vapor - 48" - 200 watt	Each	650.00
Emergency - Nickel Cadmium Battery	Each	700.00
Lead Battery	Each	360.00
Exit - Standard	Each	150.00
Exit - Low Voltage - Battery	Each	185.00
Porcelain	Each	70.00
.2 LAMPS		
Fluorescent - 48" - 40 watt	Each	6.00
48" - 60 watt	Each	7.80
96" - 75 watt	Each	10.50
Incandescent - 25 to 100 watt	Each	2.20
400 watt	Each	3.40
Metal Halide - 175 watt	Each	65.00
400 watt	Each	70.00
1000 watt	Each	145.00
Sodium - 300 watt	Each	88.00
400 watt	Each	96.00
1000 watt	Each	220.00
Add for Disposal	Each	1.00

		UNIT	COST
1606.0	**SPECIAL SYSTEMS**		
	LIGHTNING PROTECTION		
	Ground Rod and Copper Point - 3/4" to 10"	Each	85.00
	Ground Clamps - 2"	Each	105.00
	Wire - #2	LnFt	1.10
1607.0	**COMMUNICATION SYSTEMS**		
.1	TELEPHONE EQUIPMENT		
	Cabinets 24" x 36"	Each	310.00
	36" x 48"	Each	470.00
	Lines	Each	275.00
	Phones	Each	160.00
	Station	Each	580.00
.2	FIRE ALARM EQUIPMENT		
	Control Panel 10 zones	Each	1,460.00
	20 zones	Each	2,500.00
	Annunciator Panel 10 zones	Each	500.00
	Station	Each	240.00
	Horn	Each	300.00
	Heat Detector	Each	180.00
	Smoke Detector	Each	200.00
.3	SECURITY SYSTEMS		
	Standard Contact	Each	215.00
	Audible Alarm	Each	280.00
	Photo Electric Sensor - 500' Range	Each	440.00
	Monitor Panel - Standard	Each	500.00
	High Security	Each	670.00
	Vibration Sensor	Each	190.00
	Audio Sensor	Each	135.00
	Motion Sensor	Each	320.00
.4	CLOCK PROGRAM EQUIPMENT (Control Panel and Clock with Buzzer)		
	Master Time Clock 10 rm - average	Each	440.00
	20 rm - average	Each	420.00
	30 rm - average	Each	410.00
	Time Clock and Cards	Each	1,150.00
	Add for Electronic and Bells	Each	20%
.5	INTER-COMMUNICATION EQUIPMENT		
	Individual Stations	Each	135.00
	Board - Master	Each	550.00
.6	SOUND EQUIPMENT		
.61	Public Address System (Control and Speakers)		
	3 to 6 Speakers - average each speaker	Each	530.00
	7 to 9 Speakers - average each speaker	Each	480.00
	10 to 20 Speakers - average each speaker	Each	420.00
	Add for Tape / CD	Each	480.00
.62	Nurses & School Call systems (Panel & Stations)		
	10 Stations or Rooms - average each room	Each	680.00
	20 Stations or Rooms - average each room	Each	620.00
	30 Stations or Rooms - average each room	Each	540.00
	Add per Light	Each	65.00

		UNIT	COST
1607.0	**COMMUNICATIONS SYSTEMS, Cont'd...**		
.7	TELEVISION ANTENNA SYSTEMS		
	School Systems - 10 outlets	Each	325.00
	20 outlets	Each	315.00
	30 outlets	Each	295.00
	Apartment Houses - 20 outlets	Each	145.00
	30 outlets	Each	135.00
	50 outlets	Each	130.00
.8	CLOSED CIRCUIT T.V. STATIONS		
	Recorder	Each	3,600.00
	Camera	Each	1,450.00
	Surveillance	Each	1,900.00
.9	DOCTOR'S REGISTER	Each	6,600.00
	Add for Recording Register	Each	5,800.00
1608.0	**HEATING AND COOLING**		
.1	SNOW MELTING EQUIPMENT - 480 volt	SqFt	9.60
.2	ELECTRIC HEATING EQUIPMENT		
	Baseboard - 500 watt	LnFt	36.00
	750 watt	LnFt	35.00
	1000 watt	LnFt	34.00
	Cable Heating	SqFt	16.00
	Unit Heaters - 1500 watt	Each	210.00
	2500 watt	Each	300.00
1609.0	**CONTROLS AND INSTRUMENTATION**		
.1	LIGHTING CONTROL EQUIPMENT		-
.2	DIMMING EQUIPMENT		-
.3	TEMPERATURE CONTROL EQUIPMENT		-
1610.0	**TESTING**		-
1611.0	**RESIDENTIAL WORK**		
	100-Amp Service	Each	830.00
	200-Amp Service - Single Phase	Each	1,250.00
	Three Phase	Each	2,300.00
	Air Conditioner	Each	270.00
	Dimmer - 600 Watt	Each	54.00
	Dishwasher	Each	75.00
	Doorbell	Each	55.00
	Dryer - Electric	Each	105.00
	Dryer or Washer - Gas	Each	75.00
	Duplex Receptacle	Each	26.00
	Dedicated	Each	70.00
	Exhaust Fan - Bath	Each	90.00
	Kitchen	Each	95.00
	Fixture, Average	Each	90.00
	Porcelain	Each	86.00
	Furnace	Each	90.00
	Garbage Disposal	Each	75.00
	Paddle Fan	Each	196.00
	Phone - Data Box	Each	42.00
	Range Circuit	Each	130.00
	Smoke Detectors	Each	60.00
	Switches - Single	Each	26.00
	3-Way	Each	60.00
	Thermostat	Each	45.00
	Water Heater - 20 Amp 120 Volt	Each	115.00
	Whirlpool	Each	280.00

Average Square Foot Costs 2005

Housing (with Wood Frame)

Square Foot Costs do not include Garages, Land, Furnishings, Equipment, Landscaping, Financing, or Architect's Fee. Total Costs, except General Contractor's Fee, are included in each division. See Page 3 for Metropolitan Cost Variation Modifier. See Division 13 for Metal Buildings, Greenhouses, and Air-Supported Structures.

	Division	Houses - 3 Bedrooms with Basement			Houses - 3 Bedrooms without Basement		
		Low Rent	Project	Custom	Low Rent	Project	Custom
1.	General Conditions	$4.04	$4.15	$4.46	$5.25	$5.36	$5.78
2.	Site Work	2.73	3.10	3.37	4.60	4.82	4.87
3.	Concrete	3.58	3.75	3.96	6.47	6.69	7.28
4.	Masonry	4.61	4.88	6.47	6.57	6.89	8.96
5.	Steel	1.27	1.27	1.38	.63	.63	.63
6.	Carpentry	15.01	15.68	16.58	21.62	21.84	24.30
7.	Moisture & Therm. Prot.	3.58	3.58	3.80	5.03	5.19	5.62
8.	Doors, Windows & Glass	7.17	7.49	7.81	9.52	9.74	10.17
9.	Finishes	6.81	7.41	8.99	10.03	10.14	14.28
10.	Specialties	.74	.74	.80	.80	.80	1.06
11.	Equipment (Cabinets)	2.60	2.65	3.92	4.61	4.56	6.84
12.	Furnishings	.53	.53	.58	.84	.84	.84
13.	Special Construction	-	-	-	-	-	-
14.	Conveying	-	-	-	-	-	-
15.	Mechanical: Plumbing	5.26	5.32	6.27	7.84	7.84	9.69
	Heating	3.48	3.58	3.69	4.87	5.14	7.28
	Air Cond.	-	-	-	-	-	-
16.	Electrical	3.14	3.19	3.81	4.70	4.70	5.60
	SUB TOTAL	$64.55	$67.32	$75.88	$93.38	$95.16	$113.18
	CONTRACTOR'S SERVICES	5.99	6.20	7.14	8.61	8.93	10.50
	TOTAL AVERAGE Sq.Ft. Cost	$70.53	$73.51	$83.02	$101.99	$104.08	$123.68
	Unfinished Basement, Storage and Utility Areas included in Sq.Ft. Costs:				No Basement Utilities In Finished Area		
	Average Cost/ Unit	$141,064	$147,023	$207,555	$122,392	$124,898	$185,513
	Average Sq.Ft. Size/Unit	2,000	2,000	2,500	1,200	1,200	1,500
	Add Architect's Fee	$ 6,625.00	$7,420.00	$16,960.00	$5,724.00	$6,466.00	$12,720.00

	COST	
	Sq. Ft.	Ea. Unit
Add for Finishing Basement Space For Bedrooms or Amusement Rooms, 12' x 20'	$27.00	$6,400
Add for Single Garage (no Interior Finish), 12' x 20'	$25.00	$6,100
Add for Double Garage (no Interior Finish), 20 x 20'	$27.00	$10,700
Add (or Deduct) for Bedrooms, 12' x 12'	$39.00	$5,600
Add per Bathroom with 2 Fixtures, 6' x 8'	$120.00	$5,800
Add per Bathroom with 3 Fixtures, 8' x 8'	$120.00	$7,700
Add for Fireplace	-	$6,300
Add for Central Air Conditioning	-	$5,000
Add for Drain Tiling	-	$2,200
Add for Well	-	$5,000
Add for Sewage System	-	$5,500

Average Square Foot Costs
2005

Housing

Square Foot Costs do not include Land, Furnishings, Equipment, Landscaping and Financing.

		Apartments				Motels & Hotels	
	Division	2 & 3 Story	2 & 3 Story	High Rise	High Rise	1 & 2 Story	High Rise
		(1)a	(1)b	(2)	(3)	(1)b	(2)
1.	General Conditions	$ 3.89	$ 4.15	$ 5.78	$ 6.30	$ 4.25	$ 5.41
2.	Site Work	2.46	2.41	3.48	3.48	5.08	3.53
3.	Concrete	2.03	2.03	19.26	13.91	2.41	19.47
4.	Masonry	1.59	9.86	2.60	2.49	9.91	12.77
5.	Steel	1.27	1.27	2.01	2.13	1.55	1.96
6.	Carpentry and Millwork	15.68	13.16	5.04	4.70	13.10	4.09
7.	Moisture Protection	3.37	3.58	1.82	1.82	3.26	1.82
8.	Doors, Windows and Glass	5.51	5.99	6.79	7.60	6.90	8.19
9.	Finishes	10.36	10.46	10.63	10.03	10.36	14.39
10.	Specialties	.80	.80	.85	.85	1.01	1.01
11.	Equipment	1.64	2.17	1.70	1.64	2.70	2.33
12.	Furnishings	.84	.89	.95	.95	2.00	1.47
13.	Special Construction	-	-	-	-	-	-
14.	Conveying	-	2.31	3.41	3.65	-	4.10
15.	Mechanical:						
	Plumbing	6.44	6.44	6.33	6.16	7.34	9.80
	Heating and Ventilation	5.35	5.83	6.42	6.15	5.56	7.22
	Air Conditioning	1.93	1.93	2.03	1.98	2.41	4.71
	Sprinklers	-	-	2.05	1.84	1.89	1.94
16.	Electrical	6.78	6.78	7.73	8.12	9.07	9.58
	SUB TOTAL	$69.92	$80.06	$88.87	$83.76	$88.80	$113.78
	CONTRACTOR'S SERVICES	6.41	7.40	8.24	7.82	8.30	10.50
	Construction Sq.Ft. Cost	$76.33	$87.46	$97.11	$91.58	$97.10	$124.28
	Add for Architect's Fee	4.98	6.57	7.69	8.69	7.42	9.33
	TOTAL Sq.Ft. Cost	$81.31	$90.03	$104.79	$100.28	$104.52	$133.60
	Average Cost per Unit without Architect's Fee	$72,509	$83,085	$87.397	$73,266	$63,115	$80.779
	Average Cost per Unit with Architect's Fee	$77,242	$89,329	$94,314	$80,220	$67,938	$86,842
	Add for Porches – Wood - $11.00 SqFt						
	Add for Interior Garages - $10.00 SqFt						
	Average Sq.Ft. Size per Unit	950	950	900	800	650	650
	Average Sq.Ft. Living Space	750	750	750	600	450	450

(1)a Wall Bearing Wood Frame & Wood or Aluminum Facade
(1)b Wall Bearing Masonry & Wood Joists & Brick Facade
(2) Concrete Frame
(3) Post Tensioned Concrete Slabs, Sheer Walls, Window Wall Exteriors, and Drywall Partitions.

Average Square Foot Costs
2005

Commercial

Square Foot Costs do not include Land, Furnishings, and Financing.

Division	Remodeling Office Interior	Office 1-Story (1)	Office 2-Story (2)	Office Up To 5-Story (3)	High Rise (City) Above 5-Story (4)	Parking Ramp (3)
1. General Conditions	$ 2.47	$ 3.99	$ 3.99	5.93	6.25	4.41
2. Site Work & Demolition	1.50	2.03	2.03	2.03	1.87	1.34
3. Concrete	-	6.10	9.58	20.54	7.22	22.79
4. Masonry	-	10.07	9.49	14.42	3.55	1.43
5. Steel	-	-	9.37	3.39	2.42	1.67
6. Carpentry and Millwork	7.67	3.53	3.30	5.15	5.71	1.18
7. Moisture Protection	-	6.05	3.05	2.09	2.57	2.78
8. Doors, Windows and Glass	1.66	6.26	6.26	12.84	24.08	1.28
9. Finishes & Sheet Rock	10.90	11.45	11.34	13.30	14.82	1.74
10. Specialties	-	1.59	1.59	1.64	1.64	-
11. Equipment	-	4.98	4.45	3.66	4.98	-
12. Furnishings	-	1.05	1.26	1.37	1.47	-
13. Special Construction	-	-	-	-	-	-
14. Conveying	-	-	2.84	4.67	5.41	2.78
15. Mechanical:						
Plumbing	3.25	5.04	5.32	6.16	6.38	2.13
Heating and Ventilation	1.66	6.15	5.89	7.65	7.70	-
Air Conditioning	1.61	6.10	5.94	7.49	7.38	-
Sprinklers	-	1.89	1.89	2.00	2.00	-
16. Electrical	6.55	8.01	8.85	9.69	11.59	2.52
SUB TOTAL	$37.26	$84.28	$96.43	$124.02	$ 117.05	$46.05
CONTRACTOR'S SERVICES	3.31	7.88	8.93	11.55	10.82	4.31
Construction Sq.Ft. Cost	$40.57	$92.16	$105.35	$135.57	$127.87	$50.36
Add for Architect's Fee	3.71	6.78	7.63	8.90	9.54	3.60
TOTAL Sq.Ft. Costs	$44.28	$98.94	$112.98	$144.47	$137.41	$53.96
Deduct for Tenant Areas Unfinished	-	-	22.00	24.00	28.00	-
Per Car Average	-	-	-	-	-	$14,500
Add: Masonry Enclosed Ramp	-	-	-	-	-	5.50
Add: Heated Ramp	-	-	-	-	-	5.30
Add: Sprinklers for Closed Ramps	-	-	-	-	-	2.10
Deduct for Precast Concrete Ramp	-	-	-	-	-	7.60

(1) Wall Bearing Masonry and Steel Joists and Decks – No Basement
(2) Wall Bearing Masonry and Precast Concrete Decks – No Basement
(3) Concrete Frame
(4) Steel Frame and Window Wall Facade

Average Square Foot Costs
2005

Commercial and Industrial

Square Foot Costs do not include Land, Furnishings, Landscaping, and Financing.
See Division 13 for Metal Buildings

	Division	Store 1-Story (1)	Store 2-Story (2)	Production 1-Story (1)	Warehouse 1-Story (1)	Warehouse and 1-Story Office (1)	Warehouse and 2-Story Office (3)
1.	General Conditions	$ 3.68	$ 3.68	$ 3.57	$ 3.20	$ 3.31	$ 2.99
2.	Site Work & Demolition	1.87	1.44	1.66	1.61	1.93	1.55
3.	Concrete	5.46	9.95	5.56	5.35	5.94	9.10
4.	Masonry	4.98	8.53	6.63	6.68	7.47	4.03
5.	Steel	9.43	4.95	9.49	9.14	10.01	7.59
6.	Carpentry	2.69	2.74	1.74	1.68	2.35	2.07
7.	Moisture Protection	4.33	2.51	4.76	4.55	4.55	2.30
8.	Doors, Windows and Glass	5.08	4.82	4.01	2.14	4.33	3.80
9.	Finishes	5.89	5.29	5.07	1.80	3.16	4.85
10.	Specialties	1.54	1.54	1.17	1.11	1.11	1.11
11.	Equipment	1.43	1.64	1.54	-	.85	1.01
12.	Furnishings	-	-	-	-	-	-
13.	Special Construction	-	-	-	-	-	-
14.	Conveying	-	3.41	-	-	-	-
15.	Mechanical:						
	Plumbing	3.98	3.64	4.20	2.41	3.02	2.80
	Heating and Ventilation	4.01	3.96	3.10	3.21	4.28	4.87
	Air Conditioning	4.33	4.49	2.89	-	1.12	1.02
	Sprinklers	1.89	1.84	1.78	1.84	1.94	1.89
16.	Electrical	7.50	7.06	7.84	4.70	4.87	4.42
	SUB TOTAL	$68.09	$71.49	$65.00	$49.42	$60.25	$55.40
	CONTRACTOR'S SERVICES	6.62	6.62	5.99	4.52	5.57	5.04
	Construction Sq.Ft. Cost	$74.71	$78.10	$70.99	$53.93	$65.81	$60.44
	Add for Architect's Fees	5.30	4.77	4.77	3.60	4.35	4.03
	TOTAL Sq.Ft. Cost	$80.01	$82.87	$75.76	$57.53	$70.16	$64.47

(1) Wall Bearing Masonry and Steel Joists and Decks – Brick Fronts
(2) Wall Bearing Masonry and Precast Concrete Decks – Brick Fronts
(3) Precast Concrete Wall Panels, Steel Joists, and Deck

Average Square Foot Costs 2005

Educational and Religious

Square Foot Costs do not include Land, Furnishings, Landscaping and Financing.

	Division	Elementary School 1-Story (1)	High School 2-Story (1)	Vocational School 3-Story (1)	College Multi-Story (2)	Gym (3)	Dorm 3-Story (1)	Church (3)
1.	General Conditions	$ 6.35	$ 6.14	$ 6.20	$ 6.09	$ 4.99	$ 5.57	6.04
2.	Site Work	3.32	3.53	3.37	3.37	1.66	2.19	3.21
3.	Concrete	5.56	5.78	4.17	19.80	5.83	4.39	5.30
4.	Masonry	12.61	11.98	12.35	12.83	12.72	11.82	25.65
5.	Steel	9.43	11.62	11.62	3.97	11.67	9.95	1.73
6.	Carpentry	3.81	3.70	3.36	2.41	2.13	3.30	7.06
7.	Moisture Protection	4.98	3.37	3.80	3.91	5.24	4.55	5.46
8.	Doors-Windows-Glass	6.10	6.05	6.47	8.03	5.67	5.24	7.44
9.	Finishes	10.30	8.50	9.54	8.50	7.30	9.92	10.36
10.	Specialties	1.59	1.54	1.54	1.91	1.54	1.54	1.17
11.	Equipment	4.40	4.24	6.41	5.57	8.37	5.41	7.58
12.	Furnishings	-	.89	1.10	1.16	-	1.10	4.41
13.	Special Construction	-	-	-	-	-	-	-
14.	Conveying	-	-	3.15	3.15	-	3.15	-
15.	Mechanical:							
	Plumbing	7.90	8.06	8.06	10.86	5.04	7.50	5.10
	Heating-Ventilation	11.45	11.02	12.57	11.40	7.33	7.76	7.92
	Air Conditioning	-	-	-	-	-	-	6.10
	Sprinklers	1.94	1.94	1.94	1.94	1.94	1.94	
16.	Electrical	11.48	12.38	12.99	13.10	8.18	9.41	8.23
	SUB TOTAL	$101.22	$100.73	$108.65	$117.98	$89.62	$94.73	112.73
	CONTRACTOR'S SERVICES	9.35	9.35	10.08	11.03	8.30	8.82	10.50
	Constr. Sq.Ft. Cost	$110.56	$110.08	$118.73	$129.00	$97.91	$103.55	123.23
	Add Architect's Fees	7.31	7.31	7.84	8.59	6.47	6.89	8.48
	TOTAL Sq.Ft. Cost	$117.88	$117.39	$126.57	$137.59	$104.38	$110.44	$131.71

(1) Wall Bearing Masonry and Steel Joists and Decks
(2) Concrete Frame and Masonry Facade
(3) Masonry Walls and Wood Roof

Average Square Foot Costs
2005

Medical and Institutional

Square Foot Costs do not include Land, Landscaping, Furnishings, and Financing.

		Clinic 1-Story (1)	Doctors Office 1-Story (1)	Hospital 1-Story (1)	Hospital Multi-Story (2)	Nursing Home 1-Story (1)	Housing for Elderly 1-Story (1)
1.	General Conditions	$ 6.56	$ 6.67	$ 8.87	$ 8.98	$ 8.03	$ 7.35
2.	Site Work	2.09	2.03	3.21	3.26	3.42	3.53
3.	Concrete	6.79	6.74	6.90	21.94	7.12	7.22
4.	Masonry	11.66	10.49	18.97	13.25	13.52	10.97
5.	Steel	8.28	8.22	10.01	2.24	2.24	1.90
6.	Carpentry	6.66	6.50	5.71	3.70	3.98	3.81
7.	Moisture Protection	4.76	4.65	5.08	2.14	5.08	5.08
8.	Doors, Windows and Glass	7.70	8.08	10.17	6.79	8.83	7.12
9.	Finishes	10.30	10.08	13.08	12.64	12.64	10.57
10.	Specialties	1.43	1.43	1.54	1.48	1.48	1.43
11.	Equipment	7.74	7.26	11.82	11.71	5.83	2.54
12.	Furnishings	1.16	1.16	1.21	1.31	1.10	2.31
13.	Special Construction	.85	.85	1.70	1.70	.85	.85
14.	Conveying	-	-	-	6.30	-	5.88
15.	Mechanical:						
	Plumbing	10.53	11.31	17.47	16.63	11.42	11.03
	Heating and Ventilation	9.20	8.24	12.14	12.47	6.31	5.03
	Air Conditioning	7.12	6.96	12.41	10.81	7.17	2.73
	Sprinklers	1.94	1.94	1.94	1.94	2.21	1.94
16.	Electrical	13.89	11.87	21.73	19.49	12.66	11.09
	SUB TOTAL	$118.66	$114.49	$163.96	$158.79	$113.90	$102.39
	CONTRACTOR'S SERVICES	11.03	10.50	15.23	14.91	10.71	9.66
	Construction Sq.Ft. Cost	$129.69	$124.99	$179.19	$173.70	$124.61	$112.05
	Add for Architect's Fees	8.48	8.27	11.98	11.66	8.27	7.53
	TOTAL Sq.Ft. Cost	$138.17	$133.25	$191.17	$185.36	$132.88	$119.57

Average Cost to Remodel Approximately 50% Cost of New

(1) Wall Bearing Masonry and Steel Joists and Decks
(2) Concrete Frame and Masonry Façade

Average Square Foot Costs
2005

Government and Finance

Square Foot Costs do not include Land, Landscaping, Furnishings, and Financing.

		Bank & Drive-Thru 1-Story (1)	Bank & Insur. Bldg High-Rise (3)	Govt. Office Bldg Multi-Story (2)	Post Office 1-Story (1)	Court House 3-Story (2)	Jail and Prison (2)
1.	General Conditions	$ 7.25	$ 8.30	$ 8.30	$ 6.77	$ 8.19	$ 9.08
2.	Site Work	3.91	3.58	3.00	2.51	2.73	3.64
3.	Concrete	9.36	22.68	23.86	10.17	20.33	28.14
4.	Masonry	13.99	9.65	16.01	9.59	15.90	17.65
5.	Steel	4.43	2.07	2.53	3.51	2.07	5.00
6.	Carpentry	4.42	3.98	4.09	3.58	2.58	4.54
7.	Moisture Protection	3.53	4.01	4.39	3.53	4.44	5.14
8.	Doors, Windows and Glass	9.84	23.11	10.17	7.86	7.33	9.47
9.	Finishes	14.28	15.70	16.79	14.28	13.63	11.55
10.	Specialties	1.54	1.54	2.97	1.59	2.60	6.20
11.	Equipment	12.19	8.06	8.06	10.71	10.18	11.93
12.	Furnishings	1.37	1.26	-	1.84	-	-
13.	Special Construction	-	-	-	-	-	-
14.	Conveying	-	2.57	3.31	1.68	2.42	3.26
15.	Mechanical						
	Plumbing	6.38	7.56	7.39	8.01	10.14	17.47
	Heating and Ventilation	5.56	7.01	7.60	5.46	8.08	14.34
	Air Conditioning	4.55	5.14	6.53	6.53	11.45	11.98
	Sprinklers	2.05	1.94	1.94	1.94	1.94	1.94
16.	Electrical	12.77	15.40	15.90	12.10	15.23	24.64
	SUB TOTAL	$117.42	$143.55	$142.81	$111.66	$ 139.22	$185.97
	CONTRACTOR'S SERVICES	10.92	13.44	13.34	10.50	13.02	17.33
	Construction Sq.Ft. Cost	$128.34	$156.99	$156.14	$122.16	$152.24	$203.29
	Add for Architect's Fees	8.48	11.02	11.77	9.22	11.55	15.26
	TOTAL Sq.Ft. Cost	$136.82	$168.01	$167.91	$131.28	$163.79	$218.56

(1) Wall Bearing Masonry and Precast Concrete Decks
(2) Concrete Frame and Masonry Facade
(3) Concrete Frame and Window Wall

U.S. Measurements & Metric Equivalents – (For Construction)

Weights & Measures (U.S. System)

Length	1 Mile = 320 Rods = 1,760 Yards = 5,280 Feet
Area	1 Mile = 640 Acres = 102,400 Sq. Rods = 3,097,600 Sq. Yards
Area	1 Acre = 160 Sq. Rods = 4,840 Sq. Yards = 43,560 Sq. Feet
Volume	1 Cu. Yard = 27 Cu. Feet = .211 Cord
Capacity	1 Gallon = 4 Quarts = 8 Pints = .13337 Cu. Feet
Weight & Mass	1 Ton = 2,000 Pounds = 32,000 Ounces

U.S. to Metric – Area, Length, Volume & Mass Conversion Factors

Quantity	From U.S.	To Metric	Multiply By
Length	mile	km	1.609 344
	yard	m	0.914 4
	foot	m	0.304 8
	inch	mm	25.4
Area	square mile	km^2	2.590 00
	acre	m^2	4 046.87
	square yard	m^2	0.836 127
	square foot	m^2	0.092 903
	square inch	mm^2	645.16
Volume	cubic yard	m^3	0.764 555
	cubic foot	m^3	0.028 316 85
Mass-(Weight)	lb	kg	0.453
Mass-(Length)	plf	kg/m	1.488
Mass-(Square)	psf	kg/m^2	4.882
Mass-(Cube)	pcf	kg/m^3	16.018

Length Conversion Tables from Fractions to Decimals to Millimeters

Fraction	Decimal	Millimeters	Fraction	Decimal	Millimeters
1/16"	.0625	1.588	9/16"	.5625	14.288
1/8"	.125	3.175	5/8"	.625	15.875
3/16"	.1875	4.763	11/16"	.6874	17.463
1/4"	.250	6.350	¾"	.750	19.050
5/16"	.3125	7.938	13/16"	.8125	20.638
3/8"	.375	9.525	7/8"	.875	22.225
7/16"	.437	11.113	15/16"	.9375	23.813
1/2"	.500	12.700	1"	1.000	25.400

Weights – Construction Products

Substance	Wt.Lb. per Cu.Ft.	Substance	Wt.Lb. per Cu.Ft.	Substance	Wt.Lb. per Cu.Ft.
Aluminum	165	Lead	710	Concrete	144
Brass	534	Paper	58	Cement	90
Bronze	509	Glass	156	Clay	100
Copper	556	Granite	165	Sand	100
Iron	450	Limestone	160	Asphalt	81
Steel	490	Brick	140	Wood	40

Computation of Area

Square or Rectangular	=	Length x Width or Height
Triangle	=	Base x ½ Height
Circumference of Circle	=	Diameter x 3.1416
Area of Circle	=	Radius Squared x 3.1416
Cube of a Column or Cylinder	=	Radius Squared x 3.1416 x Depth
Cube of a Solid	=	Length x Width x Depth
Area of Ellipse	=	½ Length x ½ Height x 3.1416

UNITED STATES INFLATION RATES
July 1, 1970 to June 30, 2004
AC&E PUBLISHING CO. INDEX

Start July 1

YEAR	AVERAGE %	LABOR %	MATERIAL %	CUMULATIVE %
1970-71	8	8	5	0
1971-72	11	15	9	11
1972-73	7	5	10	18
1973-74	13	9	17	31
1974-75	13	10	16	44
1975-76	8	7	15	52
1976-77	10	5	15	62
1977-78	10	5	13	72
1978-79	7	6	8	79
1979-81	10	5	13	89
1980-81	10	11	9	99
1981-82	10	11	10	109
1982-83	5	9	3	114
1983-84	4	5	3	118
1984-85	3	3	3	121
1985-86	3	2	3	124
1986-87	3	3	3	127
1987-88	3	3	4	130
1988-89	4	3	4	134
1989-90	3	3	3	137
1990-91	3	3	4	140
1991-92	4	4	4	144
1992-93	4	3	5	148
1993-94	4	3	5	152
1994-95	4	3	6	156
1995-96	4	3	6	160
1996-97	3	3	3	163
1997-98	4	3	4	167
1998-99	3	3	2	170
1999-2000	3	2	3	173
2000-2001	4	4	4	177
2001-2002	3	4	3	180
2002-2003	4	5	3	184
2003-2004	3	5	2	187
2004-2005	8	4	10	195